TracerLPM (Version 1): An Excel® Workbook for Interpreting Groundwater Age Distributions from Environmental Tracer Data

By Bryant C. Jurgens, J. K. Böhlke, and Sandra M. Eberts

Techniques and Methods 4-F3

U.S. Department of the Interior
U.S. Geological Survey

U.S. Department of the Interior
KEN SALAZAR, Secretary

U.S. Geological Survey
Marcia K. McNutt, Director

U.S. Geological Survey, Reston, Virginia: 2012

For more information on the USGS—the Federal source for science about the Earth, its natural and living resources, natural hazards, and the environment, visit http://www.usgs.gov or call 1–888–ASK–USGS.

For an overview of USGS information products, including maps, imagery, and publications, visit http://www.usgs.gov/pubprod

To order this and other USGS information products, visit http://store.usgs.gov

Suggested citation:
Jurgens, B.C., Böhlke, J.K., and Eberts, S.M., 2012, TracerLPM (Version 1): An Excel® workbook for interpreting groundwater age distributions from environmental tracer data: U.S. Geological Survey Techniques and Methods Report 4-F3, 60 p.

Contents

Contents—Continued

Figures

Figures—Continued

Tables

Conversion Factors and Abbreviations and Acronyms

Conversion Factors

SI to Inch/Pound

Multiply	By	To obtain
Length		
centimeter (cm)	0.3937	inch (in.)
millimeter (mm)	0.03937	inch (in.)
meter (m)	3.281	foot (ft)
kilometer (km)	0.6214	mile (mi)
kilometer (km)	0.5400	mile, nautical (nmi)
meter (m)	1.094	yard (yd)
Area		
square meter (m^2)	0.0002471	acre
square kilometer (km^2)	247.1	acre
square kilometer (km^2)	0.3861	square mile (mi^2)
Volume		
liter (L)	33.82	ounce, fluid (fl. oz)
liter (L)	2.113	pint (pt)
liter (L)	1.057	quart (qt)
liter (L)	0.2642	gallon (gal)
cubic centimeter (cm^3)	0.06102	cubic inch (in^3)
liter (L)	61.02	cubic inch (in^3)
Mass		
gram (g)	0.03527	ounce, avoirdupois (oz)
Pressure		
kilopascal (kPa)	0.009869	atmosphere, standard (atm)
kilopascal (kPa)	0.01	bar
kilopascal (kPa)	0.2961	inch of mercury at 60°F (in Hg)
kilopascal (kPa)	0.1450	pound-force per inch (lbf/in)
kilopascal (kPa)	20.88	pound per square foot (lb/ft^2)
kilopascal (kPa)	0.1450	pound per square inch (lb/ft^2)
Density		
gram per cubic centimeter (g/cm^3)	62.4220	pound per cubic foot (lb/ft^3)

Temperature in degrees Celsius (°C) may be converted to degrees Fahrenheit (°F) as follows:

$$°F=(1.8×°C)+32$$

Temperature in degrees Fahrenheit (°F) may be converted to degrees Celsius (°C) as follows:

$$°C=(°F-32)/1.8$$

Concentrations of chemical constituents in water are given either in milligrams per liter (mg/L) or micrograms per liter (µg/L).

Conversion Factors and Abbreviations and Acronyms

Abbreviations and Acronyms

BMM	binary mixing model
BP	before 1950
CFC	chlorofluorocarbons
CO_2	Carbon dioxide
D	dispersion coefficient
DM	dispersion model
DIC	dissolved inorganic carbon
DP	dispersion parameter
EMM	exponential mixing model
EPM	exponential piston-flow model
GIS	geographic information system
IAEA	International Atomic Energy Agency
IntCal09	2009 international calibration curve
LPM	lumped parameter model
N	nitrogen
NAWQA	National Water Quality Assessment
PEM	partial exponential model
PFM	piston-flow model
pM	percent modern
pmC	percent modern carbon
pptv	parts per trillion (10^{-12}) by volume
SF_5CF_3	trifluoromethyl sulfur pentafluoride
SF_6	sulfur hexafluoride
SHcal04	2004 Southern Hemisphere calibration curve
SSW	studied supply well
TANC	Transport of Natural and Anthropogenic Contaminants
Th	thorium
TU	tritium unit
U	uranium
USA	United States of America
USGS	U. S. Geological Survey
UZ	unsaturated zone
VBA	Microsoft Visual Basic® for Applications
^3H	tritium
$^3He_{trit}$	tritiogenic helium-3
$^3H_0$	initial tritium
$^3H/^3H_0$	tritium to initial tritium ratio
^4He	helium-4
^{14}C	carbon-14
^{85}Kr	krypton-85
$\Delta^{14}C$	Delta ^{14}C value

TracerLPM (Version 1): An Excel® Workbook for Interpreting Groundwater Age Distributions from Environmental Tracer Data

By Bryant C. Jurgens, J.K. Böhlke, and Sandra M. Eberts

Abstract

TracerLPM is an interactive Excel® (2007 or later) workbook program for evaluating groundwater age distributions from environmental tracer data by using lumped parameter models (LPMs). Lumped parameter models are mathematical models of transport based on simplified aquifer geometry and flow configurations that account for effects of hydrodynamic dispersion or mixing within the aquifer, well bore, or discharge area. Five primary LPMs are included in the workbook: piston-flow model (PFM), exponential mixing model (EMM), exponential piston-flow model (EPM), partial exponential model (PEM), and dispersion model (DM). Binary mixing models (BMM) can be created by combining primary LPMs in various combinations. Travel time through the unsaturated zone can be included as an additional parameter. TracerLPM also allows users to enter age distributions determined from other methods, such as particle tracking results from numerical groundwater-flow models or from other LPMs not included in this program. Tracers of both young groundwater (anthropogenic atmospheric gases and isotopic substances indicating post-1940s recharge) and much older groundwater (carbon-14 and helium-4) can be interpreted simultaneously so that estimates of the groundwater age distribution for samples with a wide range of ages can be constrained.

TracerLPM is organized to permit a comprehensive interpretive approach consisting of hydrogeologic conceptualization, visual examination of data and models, and best-fit parameter estimation. Groundwater age distributions can be evaluated by comparing measured and modeled tracer concentrations in two ways: (1) multiple tracers analyzed simultaneously can be evaluated against each other for concordance with modeled concentrations (tracer-tracer application) or (2) tracer time-series data can be evaluated for concordance with modeled trends (tracer-time application). Groundwater-age estimates can also be obtained for samples with a single tracer measurement at one point in time; however, prior knowledge of an appropriate LPM is required because the mean age is often non-unique.

LPM output concentrations depend on model parameters and sample date. All of the LPMs have a parameter for mean age. The EPM, PEM, and DM have an additional parameter that characterizes the degree of age mixing in the sample. BMMs have a parameter for the fraction of the first component in the mixture. An LPM, together with its parameter values, provides a description of the age distribution or the fractional contribution of water for every age of recharge contained within a sample. For the PFM, the age distribution is a unit pulse at one distinct age. For the other LPMs, the age distribution can be much broader and span decades, centuries, millennia, or more. For a sample with a mixture of groundwater ages, the reported interpretation of tracer data includes the LPM name, the mean age, and the values of any other independent model parameters.

TracerLPM also can be used for simulating the responses of wells, springs, streams, or other groundwater discharge receptors to nonpoint-source contaminants that are introduced in recharge, such as nitrate. This is done by combining an LPM or user-defined age distribution with information on contaminant loading at the water table. Information on historic contaminant loading can be used to help evaluate a model's ability to match real world conditions and understand observed contaminant trends, while information on future contaminant loading scenarios can be used to forecast potential contaminant trends.

Introduction

The collection of environmental tracer data has become a routine part of groundwater-quality investigations throughout the world. Tracers provide information on groundwater ages, recharge rates, sources of recharge, mixing of groundwater masses, and groundwater-surface water interactions. The information gained from the measurement of tracers provides a more complete picture of the groundwater-flow system.

Commonly, environmental tracer measurements are made for the purpose of determining the age of groundwater. In some situations, groundwater age is estimated by relating the tracer concentration measured in a sample to the history of the tracer input in water that recharged the aquifer, assuming the tracer traveled from the recharge area to the discharge area (a well for instance) without the effects of hydrodynamic dispersion or mixing. Consequently, the water containing the tracer is assumed to have one, distinct age. This age is often referred to as a piston-flow age or an apparent age.

More often, however, water withdrawn from wells consists of many parcels of water with different ages and recharge histories because wells are typically screened across several feet or more of aquifer. Moreover, dispersion and mixing in heterogeneous aquifers can lead to broad distributions of ages even in wells with short screen intervals (Weissmann and others, 2002). Consequently, the assumption of piston-flow is unlikely to be valid for wells in many hydrogeologic settings. These processes also are applicable to investigations of groundwater discharge in springs and streams.

Since the 1950s, several mathematical models or lumped parameter models (LPMs) have been developed and applied to the interpretation of environmental tracers (Vogel, 1967; Eriksson, 1971; Maloszewski and Zuber, 1982; Amin and Campana, 1996). These models are based on simplified aquifer geometry and flow configurations that account for effects of dispersion and mixing within the aquifer or well bore. In practice, models are applied on the basis of their conceptual relevance to the aquifer and well (spring) being investigated. Although these models have been shown to explain observed tracer concentrations in different hydrogeologic settings, the practice of interpreting environmental tracers with lumped parameter models is not routine.

Recently (2011), the U.S. Geological Survey's National Water Quality Assessment (NAWQA) Program investigated the Transport of Natural and Anthropogenic Contaminants (TANC) to public-supply wells in contrasting hydrogeologic settings (Eberts and others, 2005). The first four studies to be completed were in the Central Valley aquifer in the eastern San Joaquin Valley near Modesto, California; the Floridan aquifer near Tampa, Florida; the Pomperaug River Basin near Woodbury, Connecticut; and the High Plains aquifer near Lincoln, Nebraska. The TANC study collected multiple tracers for groundwater-age determinations in each of those study areas and interpreted the tracer concentrations using lumped parameter models (LPMs). These results were compared to results from particle-tracking methods applied to detailed local-scale groundwater flow models in each study area (Eberts and others, 2012). Major findings from this study included (1) apparent (piston-flow) ages are often misleading

descriptions of the mean age of water from public-supply wells and some short-screened monitoring wells, (2) modeled contaminant responses based on age distributions can be substantially different from those based on apparent ages or mean ages, and (3) LPMs can give similar age distributions to particle tracking results for wells with water of mixed age when based on similar conceptual models and calibrated to similar tracer data. Consequently, valuable information about the age distribution of a groundwater sample from a well can be gained by evaluating LPMs with measured tracer concentrations in a relatively easy and cost-effective way compared with the development of three-dimensional groundwater-flow and transport models.

Currently available programs for estimating groundwater age from environmental tracer data include FLOWPC (Maloszewski and Zuber, 1996), BOXMODEL (Kinzelbach and others, 2002; Zoellmann and others, 2002), LUMPED and LUMPEDUS (Ozyurt and Bayari, 2002; 2005), and TRACERMODEL (Böhlke, 2006). Programs such as these are designed to work primarily with tracers of young groundwater, such as tritium (^{3}H), chlorofluorocarbons (CFCs), and sulfur hexafluoride (SF_6). Some of the programs lack either one or more LPM. BOXMODEL includes the dispersion model, but does not include the exponential piston-flow model or binary mixing models. The initial version of TRACERMODEL has the exponential piston-flow model and a binary mixing model but does not have the dispersion model. In addition, some of the programs lack a minimization algorithm to assist calibration with multiple tracers measured in a sample.

The program presented in this report, TracerLPM, is based on the workbook program TRACERMODEL (Böhlke, 2006), which has been used in several USGS research projects conducted throughout the U.S. (Böhlke and Denver, 1995; Focazio and others, 1998; Katz and others, 1999; Katz and others, 2001; Plummer and others, 2001; Böhlke, 2002; Böhlke and Krantz, 2003; Lindsey and others, 2003; Landon and others, 2008; Jurgens and others, 2008; Katz and others, 2009; Brown and others, 2009). The TRACERMODEL workbook emphasizes simultaneous modeling of multiple tracers to evaluate the mean age and age distribution of a sample and features graphical analysis of "tracer-tracer" plots. An additional workbook version of TRACERMODEL was developed for comparing tracer time series data and models. Many refinements and improvements in TRACERMODEL, as well as inclusion of additional binary models, tracers of old groundwater (carbon-14, helium-4), a minimization scheme, and forecasting of tracer concentrations, have been implemented in a new workbook program that is presented herein—TracerLPM.

The purpose of this report is to document the Excel® workbook program, TracerLPM, which is intended for use in evaluating groundwater age distributions and mean ages in samples from wells and springs on the basis of environmental tracers in groundwater. This report is organized into four parts. The first section describes the models and their applicability to certain hydrogeologic settings. The second section describes the environmental tracers currently included in this workbook. The third section gives an overview of the workbook and a detailed description of the individual worksheets. The fourth section contains examples illustrating the use and functionality of the program.

Lumped Parameter Model Calculations

The TracerLPM workbook contains five LPMs that can be used to determine the age distribution and mean age for a sample: piston-flow model (PFM), exponential mixing model (EMM), exponential piston-flow model (EPM), partial exponential model (PEM), and dispersion model (DM). Each of these models can be combined with another model to create a binary mixing model (BMM). This leads to 25 possible BMM combinations. The LPMs correspond to different configurations of groundwater flow from an inlet position in the aquifer (recharge area) to an outlet position in the aquifer (a well or spring), and are represented mathematically as transit-time distribution functions or exit-age distribution functions [$g(t)$] (Maloszewski and Zuber, 1982). A water sample is envisioned to consist of many "parcels" that followed different flow paths to the sample site; each parcel represents a relatively discrete groundwater age and tracer concentration. For a steady state groundwater system, simulated tracer concentrations at an outlet position in the aquifer can be calculated from the tracer input history at the inlet position of the aquifer using the exit age distribution function and the decay function for the tracer (e.g. radioactive decay or biodegradation):

$$C_{out}(t) = \int_{-\infty}^{t} C_{in}(t')e^{-\lambda(t-t')} g(t-t')dt' \qquad (1)$$

where

$\quad C_{out}(t)$ is outlet tracer concentration,
$\quad C_{in}(t')$ is concentration of tracer at inlet at time t',
$\qquad t$ is sample date,
$\qquad t'$ is date at which a water parcel entered the system,
$\qquad \lambda$ is decay constant, fractional loss per unit of time, and
$\quad t - t'$ is age of water parcel.

The lumped parameter model approach assumes tracers are injected and detected in the fluid flux and that the tracers behave conservatively (except for possible radioactive decay or degradation) and travel with the water. Therefore, the mean age inferred by tracer concentrations is equal to the mean age of water discharging from the system or to a screened interval. For some tracers or hydrogeologic situations, deviations between the tracer travel time and the water travel time can lead to unreliable estimates of the age distribution and mean age of water in a sample. It is, therefore, important to consider how processes such as diffusion and geochemical exchange can affect certain tracers that are not part of the water molecule and lead to large discrepancies between tracer-based ages and water ages.

The mean age of a sample (τ_s) is derived from the exit-age distribution function that describes the tracer concentrations in the sample:

$$\tau_s = \int_{-\infty}^{t} (t-t')g(t-t')dt \qquad (2)$$

which can be approximated numerically by:

$$\tau_s = \sum_{i=1}^{\infty} t_i X_i (\Delta t) \qquad (3)$$

where

$\quad t_i$ is age of water parcel $(t-t')$,
$\quad X_i$ is fraction of the sample represented by a water parcel, corresponding to a given age increment, and
$\quad \Delta t$ is time step (age increment).

Commonly, the mean age of the sample, τ_s, is not equal to the mean age of groundwater in the whole aquifer, τ_{aq}, but in some cases, such as in the EMM or EPM, the mean age of a sample also corresponds to the mean age of groundwater in the whole aquifer.

Closed form analytical solutions to equation 1 were derived for each LPM age-distribution function, $g(t-t')$, and were implemented in the workbook by using an Excel add-in called "TracerLPMfunctions" (compiled for 32-bit and 64-bit versions of Excel). The XLL add-in contains worksheet functions of the LPMs, programmed in the C++ language using Microsoft® Visual Studio® and the Microsoft Office Excel® 2010 Software Development Kit (see Compatibility section in appendix B for more details).

Piston-Flow Model (PFM)

The piston-flow model (PFM) assumes a tracer travels from the inlet position (recharge area) to the outlet position (a well or spring, for example) without hydrodynamic dispersion or mixing. The PFM can be applicable to hydrogeologic settings where dispersion is low, average linear velocity is high, or the flow path from recharge to discharge is short. Tracers measured from shallow, short-screened monitoring wells in unconfined aquifers or short-screened wells in confined aquifers with a small recharge area can follow piston-flow behavior approximately (fig. 1). The exit-age distribution function of the piston-flow model is as follows:

$$\text{PFM}_{g(t-t')} = \delta(t - t' - \tau_s) \tag{4}$$

where

δ is Dirac delta function.

The PFM is calculated by using the following formula:

$$C_{out}(t) = C_{in}(t - \tau_s)e^{(-\lambda - \tau_s)} \tag{5}$$

$$\text{for } t = \tau_s; \ 0 \text{ for } t \neq \tau_s$$

Piston-flow Model (PFM)

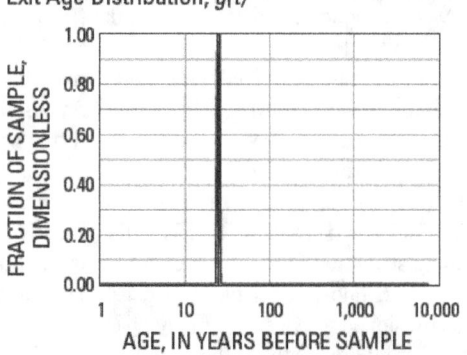

Exit Age Distribution, *g(t)*

Figure 1.   Schematic diagram of idealized hydrogeologic aquifer configurations in which the piston-flow model could be applicable. The top diagram shows an unconfined aquifer receiving areal recharge with short-screened wells located in the shallow part of the aquifer. The second diagram (modified from Maloszewski and Zuber, 1982) shows a confined aquifer with a small recharge area and a well located down-gradient of the recharge area. Dispersion or diffusion processes are assumed to have little or no effect on tracer concentration gradients. Right-pointed arrows are sample points. The bottom graph shows the exit-age frequency distribution expected from piston-flow transport of the tracer from the recharge area to the well having a mean age of 25 years, in 1-year increments.

Exponential Mixing Model (EMM)

The exponential mixing model (EMM) is applicable to homogeneous, unconfined aquifers of constant thickness receiving uniform recharge (fig. 2). This situation leads to vertical stratification of groundwater age, which increases logarithmically from zero at the water table to ages that approach infinity at the base of the aquifer (Vogel, 1967; Appelo and Postma, 1996). This model can be appropriate for fully penetrating wells or aquifers that discharge to springs or streams. The EMM model applies where longitudinal and transverse dispersion does not occur along flowlines and mixing occurs within the well bore or spring rather than in the aquifer. The EMM also describes the age distribution of a completely mixed reservoir (Eriksson, 1971). The EMM exit-age distribution function is as follows:

$$EMM_{g(t-t')} = \frac{1}{\tau_s} e^{\left(\frac{t-t'}{\tau_s}\right)} \qquad (6)$$

The EMM is calculated by using the following closed form solution of the convolution integral for each age increment (Δt) starting from the sample date minus the age increment and stepping backwards in time until the output concentration does not change by more than 10^{-6}:

$$C_{out}(t) = \sum_{t'=-\infty}^{t-\Delta t} C_{in}(t') \frac{1}{\tau_s} \frac{1}{\frac{1}{\tau_s} + \lambda}$$
$$\left[e^{\left[-\left(\frac{1}{\tau_s}+\lambda\right)(t-t'-\Delta t)\right]} - e^{\left[-\left(\frac{1}{\tau_s}+\lambda\right)(t-t')\right]} \right] \qquad (7)$$

Exponential Mixing Model (EMM)

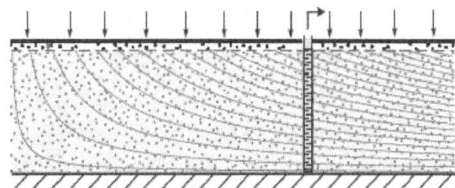

Exit Age Distribution, *g(t)*

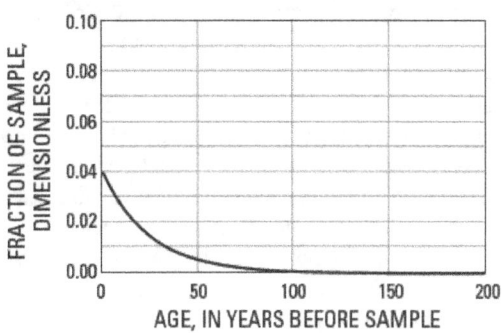

Figure 2. Schematic diagram of idealized aquifer configuration in which the exponential model could be applicable. The top diagram shows an unconfined aquifer receiving uniform recharge with samples taken from a well screened across the entire thickness of the aquifer or a spring. Mixing is assumed to occur in the well or spring and not in the aquifer. Right-pointed arrows are sample points. The bottom graph shows the exit-age frequency distribution expected from exponential mixing of the tracer in the well or spring having a mean age of 25 years, in 1-year increments.

Exponential Piston-Flow Model (EPM)

The exponential piston-flow model (EPM) can be used to describe an aquifer that has two segments of flow in series: a segment of exponential flow followed by a segment of piston-flow. This model can be used to describe discharge from an aquifer of constant thickness with an upgradient unconfined portion receiving areally distributed recharge (the exponential part) connected to a downgradient confined portion or an unconfined portion receiving little to no recharge (piston-flow part; fig. 3). The EPM also can be used to describe piston-flow transport within the unsaturated zone followed by exponential mixing; however, piston-flow transport through the unsaturated zone has been implemented separately for all models in TracerLPM. In addition, some tracers, like tritiogenic helium-3 for example, cannot be modeled correctly using an EPM to describe unsaturated zone (UZ) transport. Therefore, the EPM, as implemented in this program, is intended for situations where exponential flow precedes piston-flow within the saturated zone. The EPM age distribution function is as follows:

$$\mathrm{EPM}_{g(t-t')} = \frac{n}{\tau_s} e^{\left[-\frac{n(t-t')}{\tau_s}+n-1\right]}, \text{ for } t \geq \tau_s\left(1-\frac{1}{n}\right); 0 \tag{8}$$

where

$$n \text{ is } \frac{\text{total volume}}{\text{exponential volume}} = \frac{x^*+x}{x} = \frac{x^*}{x}+1$$
$$= \mathrm{EPM\,ratio}+1$$

The EPM has two parameters: mean age and the EPM ratio. The EPM ratio is the ratio of the length of area at the water table not receiving recharge (x^*) to the length of area receiving recharge (x), or the ratio of the piston-flow and exponential model, $\frac{x^*}{x}$, components (Cook and Böhlke, 2000; Böhlke, 2006). This ratio is used to calculate the parameter, n, defined in Maloszewski and Zuber (1982) as the ratio of the total volume to the volume with the exponential distribution $\left(n = \frac{x^*+x}{x}\right)$. The EPM ratio causes the age distribution to vary from completely exponential (EPM ratio equals 0, n equals 1) to nearly piston-flow (EPM ratio greater than 5).

EPM tracer concentrations are caluculated by using the following closed form solution of the convolution integral for each age increment (Δt) starting from the youngest parcel and stepping backwards in time until the output concentration does not change by more than 10^{-6}:

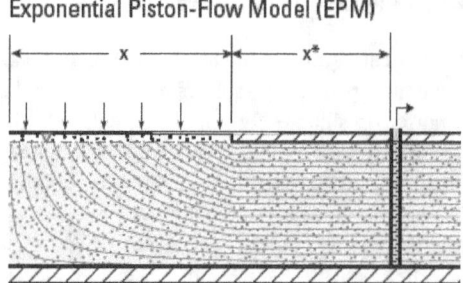

Exponential Piston-Flow Model (EPM)

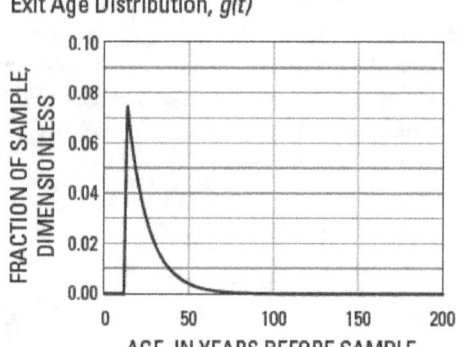

Exit Age Distribution, $g(t)$

Figure 3. Schematic diagram of idealized aquifer configuration in which the exponential piston-flow model could be applicable. The top diagram shows a partly confined aquifer with a recharge area of length x and a confined part of length x^*. Right-pointed arrows are sample points. The bottom graph shows the exit-age frequency distribution expected from exponential piston-flow transport of the tracer from the recharge area to the well having an EPM ratio of 1 and mean age of 25 years, in 1-year increments.

$$C_{out}(t) = \sum_{t'=-\infty}^{t-\Delta t} C_{in}(t') \frac{n}{\tau_s} \frac{1}{\frac{n}{\tau_s}+\lambda} \tag{9}$$

$$\left[e^{\left[-\left(\frac{n}{\tau_s}+\lambda\right)(t-t'-\Delta t)+n-1\right]} - e^{\left[-\left(\frac{n}{\tau_s}+\lambda\right)(t-t')+n-1\right]} \right]$$

$$\text{for } t \geq \tau_s\left(1-\frac{1}{n}\right); 0 \text{ for } t < \tau_s\left(1-\frac{1}{n}\right)$$

Partial Exponential Model (PEM)

The partial exponential model (PEM) is applicable to the same type of aquifer as the exponential mixing model, but is used when only the lower part of the aquifer is sampled (fig. 4). Public supply wells are commonly constructed this way. The PEM presented here is a special case of the more general PEM described in appendix A. The PEM exit-age distribution function is as follows:

$$\text{PEM}_{g(t-t')} = \frac{n}{\tau_{aq}} e^{\left(\frac{t-t'}{\tau_{aq}}\right)}, \text{ for } t - t' \geq \tau_{aq} \ln(n); 0 \qquad (10)$$

where

$$n \text{ is } \frac{\text{total volume}}{\text{sampled volume}} = \frac{z^* + z}{z} = \frac{z^*}{z} + 1 = \text{PEM ratio} + 1$$

$$\tau_{aq} = \frac{\tau_s}{\left(1 - \ln\left(1 - \frac{z^*}{z + z^*}\right)\right)}$$

The PEM has two parameters: mean age of the sampled interval, τ_s, and the PEM ratio. The PEM ratio is defined as the ratio of the unsampled thickness of the aquifer to the sampled thickness $\left(\frac{z^*}{z}\right)$ and is used to calculate the parameter n in the above equation, which is the ratio of the total thickness to the sampled thickness. For wells screened across the entire saturated thickness of the aquifer (PEM ratio equals 0, n equals 1), the EMM is obtained, while wells with small screened intervals are more similar to the PFM. This definition makes the PEM appealing to use for long-screened wells that do not sample the upper portion of the aquifer because the PEM ratio can be estimated independently from well construction and depth-to-water information. In some cases, the bottom of the screened interval can be used when the aquifer depth is unknown. However, in situations where the bottom of the screened interval is far from the bottom of the aquifer, this assumption can be invalid (see appendix A).

PEM tracer concentrations are calculated using the following closed form solution of the convolution integral for each age increment (Δt) starting from the sample date minus the age increment and stepping backwards in time until the output concentration does not change by more than 10^{-6}:

Partial Exponential Model (PEM)

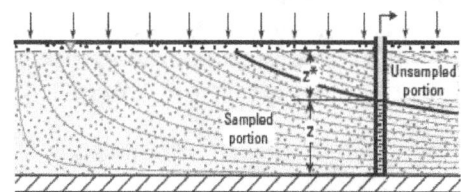

Exit Age Distribution, g(t)

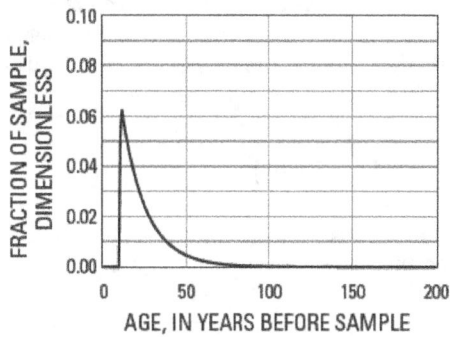

Figure 4. Schematic diagram of idealized aquifer configuration in which the partial exponential model could be applicable. The top diagram shows an unconfined aquifer receiving uniform recharge with a partially screened well of length (z) beginning at some depth below the water table (z^*). Mixing occurs in the well or spring and not in the aquifer. Right-pointed arrows are sample points. The bottom graph shows the exit-age frequency distribution expected from a well of this type having a PEM ratio of 1 and mean age of 25 years (in 1-year increments).

$$C_{out}(t) = \sum_{t'=-\infty}^{t-\Delta t} C_{in}(t') \frac{n}{\tau_{aq}} \frac{1}{\frac{1}{\tau_{aq}} + \lambda} \qquad (11)$$

$$\left[e^{\left[-\left(\frac{1}{\tau_{aq}}+\lambda\right)(t-t'-\Delta t)\right]} - e^{\left[-\left(\frac{1}{\tau_{aq}}+\lambda\right)(t-t')\right]} \right]$$

$$\text{for } t - t' \geq \tau_{aq} \ln(n); 0$$

As is evident from the exit-age distribution, the PEM is similar to the EPM. In fact, the PEM, as implemented in the initial version of TracerLPM, will give the same age distribution and tracer output concentrations as the EPM when τ_{PEM} is equal to τ_{EPM} and the following is true:

$$\text{EPM ratio} = \ln(\text{PEM ratio} + 1) \qquad (12)$$

Although the models can be parameterized to give identical age distributions, they are presented differently to facilitate evaluation of different hydrogeologic situations. This is especially useful where dimensions x or z are available from field data and can be compared to the models.

Dispersion Model (DM)

The dispersion model is based on a solution to a one-dimensional advection dispersion equation for a semi-infinite medium with an instantaneous injection and detection of the tracer in the fluid flux (Kreft and Zuber, 1978; Maloszewski and Zuber, 1982). The dispersion model can give an approximate description of age distributions in samples from many aquifer configurations (fig. 5). The dispersion model has two parameters, which are mean age and the dispersion parameter:

$$DM_{g(t-t')} = \frac{1}{\tau_s} \frac{1}{\sqrt{4\pi DP \frac{t-t'}{\tau_s}}} e^{\left(-\frac{\left(1-\frac{t-t'}{\tau_s}\right)^2}{4DP\frac{t-t'}{\tau_s}}\right)} \qquad (13)$$

where

$$DP = \text{dispersion parameter} = \frac{\text{Dispersion coefficient}(D)}{vx}$$

The dispersion parameter (DP) is the inverse of the Peclet number or the ratio of the dispersion coefficient (D) to the velocity (v) and outlet position (x). In practice, the dispersion parameter describes the relative width and height of the age distribution and is mainly a measure of the relative importance of dispersion (mixing) to advection (Zuber and Maloszewski, 2001). Consequently, increasing values of the dispersion parameter tend to move the peak toward younger age parcels and decreasing values of the dispersion parameter produce more narrow age distributions with taller peaks centered on the mean age (Maloszewski and Zuber, 1996).

The dispersion model has the following closed form solution (Maloszewski and Zuber, 1996; Kinzelbach and others, 2002) and is calculated by numerically integrating the following equation for each age increment (Δt) starting

Dispersion Model (DM)

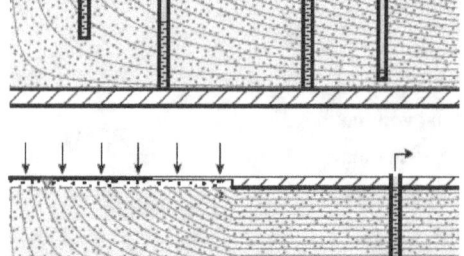

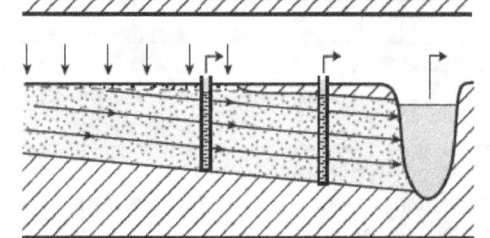

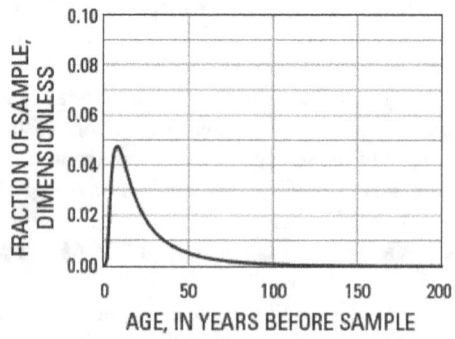

Exit Age Distribution, *g(t)*

Figure 5. Schematic diagrams of idealized aquifer configurations in which the dispersion model could be applicable. Mixing occurs within the aquifer because of variations in groundwater velocity related to heterogeneity on various spatial scales, and can therefore disturb the age distributions of some of the other conceptual models (modified in part from Maloszewski and Zuber (1982). Right-pointed arrows are sample points. The bottom graph shows the exit age frequency distribution expected from the dispersion model with a dispersion parameter of 0.5 and mean age of 25 years, in 1-year increments.

from the youngest age of water in the sample and stepping backwards in time until the output concentration does not change by more than 10^{-7}. Numerical integration is performed using adaptive quadrature (Burden and Faires, 2005).

$$C_{out}(t) = \frac{1}{\sqrt{4\pi DP}} e^{\left(\frac{1}{2DP}\right)} \sum_{t'=-\infty}^{t-\Delta t} C_{in}(t') \int_{x_2}^{x_1} x^{-\frac{3}{2}} e^{\left(-ax-\frac{b}{x}\right)} dx \quad (14)$$

where

$$a = \lambda \tau_s + \frac{1}{4DP},$$

$$b = \frac{1}{4DP},$$

$$x = \frac{t-t'}{\tau_s},$$

$$x_1 = \frac{t-t'}{\tau_s},$$

$$x_2 = \frac{t-t'-\Delta t}{\tau_s}.$$

Binary Mixing Model (BMM)

Binary mixing models (BMMs) describe two-component mixtures in which each component can be described by one of the models given above. Binary mixing models can be appropriate for wells screened across multiple aquifer units and aquifers with short-circuit pathways that result in age mixtures of significantly different mean ages (fig. 6). BMMs also have been used to describe tracer concentrations in karstic aquifers and watersheds with transmissivity contrasts (Maloszewski and others, 1983; Michel, 2004; Long and Putnam, 2006; Katz and others, 2008). BMMs have the prefix "BMM" and are followed by two models that define the two components of water in the mixture. For example, a "BMM-PFM-EMM" model describes a binary mixture in which one component of the mixture is modeled by using piston-flow and the other component is modeled by using an exponential mixing model. The mixing fraction, f_1, corresponds to the first model listed in the composite model name. Binary mixing models are calculated by using the following equation:

$$C_{out} = f_1 C_1 + (1-f_1) C_2 \quad (15)$$

where

C_1 is concentration of tracer in first component,
C_2 is concentration of tracer in second component, and
f_1 is fraction of first component in the overall binary mixture.

Binary Mixing Model (BMM)

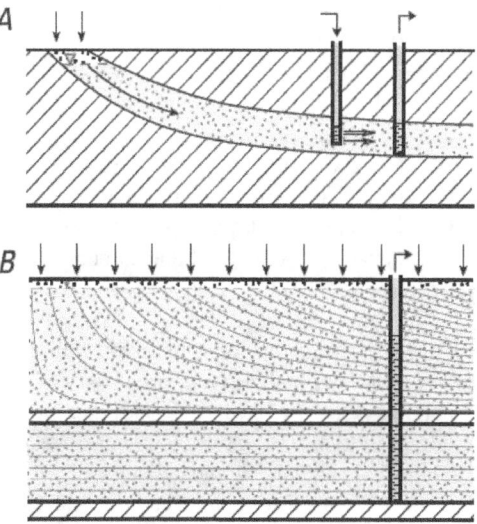

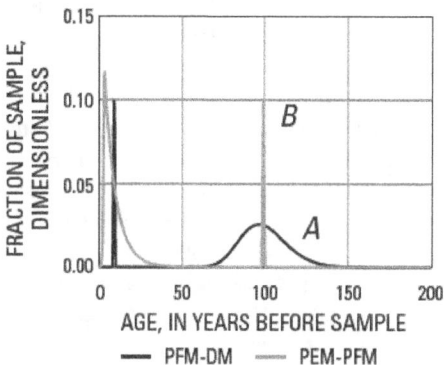
Exit Age Distribution, g(t)

Figure 6. Schematic diagram of idealized aquifer configurations in which the binary mixing model could be applicable. Right-pointed arrows are sample points and down pointed arrows are short-circuit pathways to the aquifer. The bottom graph shows the exit age frequency distribution expected from two binary mixing models. The solid line A is a BMM composed of a young, PFM component and an older, DM component (BMM-PFM-DM). The PFM simulates a short-circuit pathway caused by an abandoned upgradient well and the DM simulates the natural flow of groundwater from the recharge area to the pumping well in the top figure. The age distribution shows a PFM component age of 10 years and a fraction of 10 percent along with a DM component having a mean age of 100 years and a dispersion parameter 0.01. The top diagram was modified from Maloszewski and Zuber (1982). The dotted line B is a BMM composed of a young PEM component and an older PFM component (BMM-PEM-PFM). The PEM simulates the groundwater from the upper unconfined portion of the aquifer and the PFM simulates the contribution of old water below the confining unit. The PEM has a mean age of 10 and a PEM ratio of 0.5, whereas the PFM has a mean age of 100 years. The BMM of B is a mixture in which 90 percent is from the younger PEM component and 10 percent is from the older PFM component.

Environmental Tracers

TracerLPM can simulate outlet concentrations for tracers in groundwater with a wide range of ages (less than 1 year to more than 10,000 years). Tracers of groundwater commonly refer to the atmospheric environmental tracers most often used to date groundwater on time scales of years to decades. These tracers include tritium (^{3}H), tritiogenic helium-3 (^{3}He$_{trit}$), initial tritium (^{3}H$_o$), tritium to initial tritium ratio (^{3}H/^{3}H$_o$), krypton-85 (^{85}Kr), chlorofluorocarbon-11 (CFC-11), chlorofluorocarbon-12 (CFC-12), chlorofluorocarbon-13 (CFC-13), chlorofluorocarbon-113 (CFC-113), sulfur hexafluoride (SF$_6$), trifluoromethyl sulfur pentafluoride (SF$_5$CF$_3$), and nuclear bomb derived carbon-14 (^{14}C). Tracers of old groundwater are used to date groundwater on time scales of the order of 100 to tens of thousands of years before present, and include natural ^{14}C and helium-4 (^{4}He).

Most of the tracers listed above require some degree of interpretation of the analytical results, and all data should be evaluated for laboratory and field quality-assurance purposes prior to their entry into TracerLPM. Only ^{3}H and, in some instances, ^{14}C values can be entered into the workbook without prior manipulations to the analytical results. In many cases, ^{14}C data can be adjusted for geochemical reactions by using other programs, such as NETPATH, NetpathXL, or NETPATH-WIN (Plummer and others, 1994; Parkhurst and Charlton, 2008, El-Kadi and others, 2011). Uncorrected ^{14}C measurements, however, often can be simulated properly if the dilution factor happens to be known for a set of data and is applied to the ^{14}C input curve using the scaling factor on the *TracerInput* worksheet, but the scaling factor is often only known through inverse geochemical modeling programs.

Most of the gas tracers used to date groundwater require adjustments to the analytical results to account for recharge temperature, recharge elevation and excess air (Cook and Herczeg, 2000; Kazemi and others, 2006). CFCs, SF$_6$, and SF$_5$CF$_3$ concentrations in this program are expressed as the atmospheric gas-mixing ratios in parts per trillion by volume (pptv) that would be in equilibrium with the measured concentrations in the water at the estimated recharge temperature, adjusted for the effect of recharge elevation. The normalized gas-mixing ratios are preferred to the measured concentrations because the normalization removes effects of local variations in recharge temperature and elevation, and permits direct comparison of the data to regional records of atmospheric-mixing ratios. In addition, contamination, sorption, or degradation can significantly affect the concentrations of CFCs measured in groundwater (International Atomic Energy Agency, 2006). These factors should be considered when evaluating the suitability of the CFC data for age determination. The determination of tritiogenic helium-3 (^{3}He$_{trit}$) also can require an estimation of the ^{3}He/^{4}He ratio of terrigenic helium if terrigenic helium concentrations compose a significant amount of the total

helium in the sample (Schlosser and others, 1989). This workbook can help elucidate problems where the measured tracer concentrations cannot be explained by the LPMs.

Tracers of Young Groundwater

This workbook contains the input history for several of the anthropogenic atmospheric environmental tracers listed above. For CFC-11, CFC-12, CFC-13, CFC-113, SF$_6$, and SF$_5$CF$_3$, the atmospheric input histories for northern and southern Hemispheres were compiled by the U.S. Geological Survey Chlorofluorocarbon Laboratory at half-year time increments. These data were interpolated to monthly data, and the interpolated value at the mid-point of each month was used as the average value for the month. For example, the interpolated value calculated at February 15, 2000, (2000.12 in decimal years) was entered as the value at February 1, 2000 (2000.083 in decimal years). For this workbook, integration of the convolution equation (1) is performed on a monthly basis. Therefore, the input history entered into the workbook should be the average value for the month, if monthly data are available. This distributes the error of integration evenly before and after the mid-point of the month. Consequently, users should be aware that tracer input data can need to be offset before being entered into this workbook.

This workbook also includes data for ^{14}C in the troposphere for three northern Hemisphere zones and the southern Hemisphere (Hua and Barbetti, 2004) that resulted from atmospheric bomb testing since 1955, and is described more fully in the following "Carbon-14" section. These data do not account for exchanges among carbon reservoirs in the soil or geochemical reactions in the saturated zone. Hence, the concentration of ^{14}C in recharge can differ significantly from the concentration reported from these data. Geochemical inverse modeling can be required to determine whether ^{14}C in groundwater has been diluted by ^{14}C-dead carbon sources (Plummer and others, 1994). The areas and boundaries of the northern Hemisphere zones and the southern Hemisphere can be found in Hua and Barbetti (2004) and also at the CaliBomb (Reimer and others, 2004) web page (last accessed 11/2011): http://calib.qub.ac.uk/CALIBomb/frameset.html

This workbook includes ^{3}H in precipitation data for Ottawa, Canada, and Vienna, Austria. The Ottawa record was obtained from the International Atomic Energy Agency (IAEA) web site. For periods of missing data, the ^{3}H record in precipitation was reconstructed from the Chicago (Midway), Illinois, precipitation station using the correlation coefficients developed from the periods of overlap (Michel, 1989). The Vienna record was reconstructed from tritium in precipitation data by using modified procedures developed by Doney and others (1992; Luis Araguas, IAEA, written commun., 2011). The background (pre-bomb) concentrations of tritium in precipitation for Ottawa and Vienna was estimated to be 8 and 4 tritium units (TU), respectively. For periods beyond the available tritium records (2007 for Ottawa and 2001 for Vienna), tritium concentrations were extrapolated on the basis of the tritium trend in precipitation by using least squares regression to estimate an exponential decline in ^{3}H until the year 2020. These trends could need to be modified as ^{3}H in precipitation approaches pre-bomb values. Tritium in precipitation records for local or regional studies in the USA can be derived from nearby IAEA precipitation stations or from the ^{3}H data set compiled by Michel (1989). Only ^{3}H need be entered to simulate the concentrations of the other tritiogenic tracers, ^{3}He$_{trit}$, ^{3}H$_o$, and ^{3}H/^{3}H$_o$. The input function for ^{3}He is calculated from the ^{3}He$_{trit}$ input curve for every age in a sample:

$$C_{in}(^3\text{He}_{trit}) = C_{in}(^3\text{H})\left(1 - e^{\left[-\lambda(t-t')\right]}\right) \qquad (16)$$

For this program, the decay product, ^{3}He$_{trit}$, and ^{3}H$_o$ are modeled with the assumption that ^{3}He$_{trit}$ moves with ^{3}H in the water. The ^{3}H/^{3}H$_o$ ratio is calculated subsequently from the modeled ^{3}H and ^{3}H$_o$ concentrations in the sample.

Carbon-14

Radioactive decay of natural ^{14}C (half-life of 5,730 years) commonly is used to date groundwater on timescales of the order of 100 to tens of thousands of years, and this capability is included in TracerLPM. The concentration of ^{14}C in the atmosphere has been lower, and more often much higher, than 100 percent modern carbon (pmC) over the last 50,000 years, mainly as a result of natural variations in the Earth's geomagnetic field (Stuiver, 1961; 1965) and more recently from above-ground nuclear weapons testing (Clark and Fritz, 1997). Because of these fluctuations, a measured ^{14}C concentration in groundwater could correspond to several ages in the past. These observations have been long recognized in the radiocarbon community (Stuiver 1982, Aitchison and others, 1989), and radiocarbon calibration programs have been developed to determine a range of ages that could be equally valid for a single ^{14}C concentration and measurement error (www.radiocarbon.org).

Calibration curves also can be used with lumped parameter models for purposes of determining ^{14}C concentrations in groundwater mixtures. For TracerLPM, ^{14}C input curves for the Northern Hemisphere were constructed by combining the international calibration curve, IntCal09 (Reimer and others, 2009), with modern tropospheric ^{14}C data for three Northern Hemisphere zones (zones 1, 2, or 3; Hua and Barbetti, 2004). Likewise, an input curve for the Southern Hemisphere was constructed by combining the Southern Hemisphere calibration curve, SHcal04 (McCormac and others, 2004), and modern tropospheric ^{14}C data for the Southern Hemisphere (Hua and Barbetti, 2004). Records for IntCal09, SHcal04, and tropospheric ^{14}C data were reported

as Delta ^{14}C values (Δ^{14}C). These values were converted to absolute percent modern (pM) by using the following relation (Stuiver and Pollach, 1977):

$$pM = \left(\frac{\Delta^{14}C}{1000} + 1\right)100 \qquad (17)$$

The period of record for tropospheric ^{14}C in the Northern Hemisphere data sets (zones 1, 2, and 3) began in year 1955.5 and ended in year 1999.5 (Hua and Barbetti, 2004). For the Southern Hemisphere, data reported by Hua and Barbetti (2004) began in year 1955 and ended in year 2001. In TracerLPM, for all four sets of data, ^{14}C values between 1950 and the first year of tropospheric data were estimated from an exponential interpolation. For the period after the end of tropospheric data until the year 2020, least squares regression was used to estimate an exponential decline in ^{14}C for the years after 1999.5 until the year 2020.

The Southern Hemisphere long-term calibration curve extends back to 11,000 years before 1950 (BP), while the Northern Hemisphere IntCal09 curve extends back to 50,000 years BP. Because the concentration of ^{14}C in the Southern Hemisphere closely follows the concentration in the Northern Hemisphere during the last 11,000 years BP, a relationship or scaling factor can be found that approximately relates the two curves (fig. 7). The scaling factor was determined by multiplying one curve by an initial scaling factor and adjusting the scaling factor until the error between the two curves was minimized. This method assumes the ^{14}C concentration is merely diluted or more concentrated in the Southern Hemisphere and not shifted in time.

This procedure resulted in a scaling factor of 0.9930, which indicates the ^{14}C concentration has been slightly lower in the Southern Hemisphere than in the Northern Hemisphere over the last 11,000 years BP. Past research has shown that atmospheric ^{14}C concentrations in the Southern Hemisphere are diluted in comparison with the Northern Hemisphere because a larger areal expanse of ocean water and slightly higher wind speeds cause more ^{14}C depleted carbon dioxide (CO_2) from the ocean to enter the southern atmosphere (McCormac and others, 1998; 2004).

The scaling factor was used to construct the atmospheric ^{14}C concentration for the Southern Hemisphere from 11,000 to 50,000 years BP (dashed orange-yellow line on fig. 7) from the IntCal09 curve (fig. 7). McCormac and others (2004) do not recommend extending the Southern Hemisphere curve beyond 11,000 years because of increased uncertainty created by larger-scale carbon reservoir changes that could have altered the interhemispheric offset used to calibrate the curve to 11,000 years. Consequently, application of the Southern Hemisphere curve for samples with ages older than 11,000 years might not be reliable until more data are available to extend the curve for older atmospheric ^{14}C.

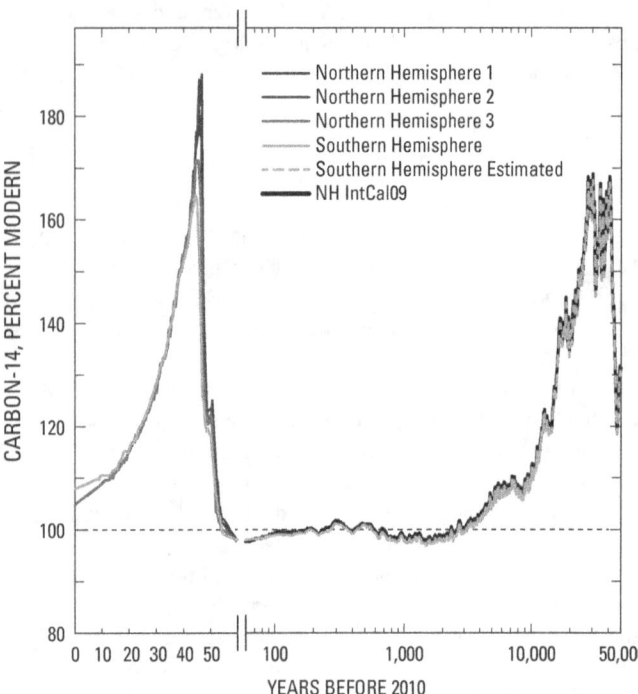

Figure 7. Carbon-14 concentration in the atmosphere for the last 50,000 years.

Nevertheless, uncertainties in the hemispheric ^{14}C gradients are likely to be small compared to other uncertainties in the evaluation of ^{14}C data in groundwater with TracerLPM.

The development of the ^{14}C input curves for long time scales allows consistency between the ages determined with lumped parameter models and the piston-flow ages determined from radiocarbon calibration programs (www.radiocarbon.org). For determining piston-flow ^{14}C ages, the radiocarbon calibration programs can be the preferred method. However, mixtures of water of different ages, and especially those having both old, naturally produced ^{14}C and nuclear bomb derived ^{14}C, can be evaluated more appropriately using this program.

For TracerLPM, the measured ^{14}C concentration for samples should be entered as absolute percent modern carbon (pmC) and not absolute percent modern (pM). The atmospheric ^{14}C concentrations used in TracerLPM are expressed as pM, and have been normalized for carbon isotope fractionation. It is standard practice in the radiocarbon community to normalize the measured ^{14}C content so that samples of different chemical composition (carbonate rock, plants) but having carbon of the same age are comparable. Groundwater ^{14}C measurements also are commonly reported in pM after conventional adjustment for carbon isotope fractionation. However, ^{14}C data for groundwater samples should not be normalized for δ^{13}C isotope fractionation because δ^{13}C variations in groundwater are mainly a result of geochemical reactions that took place in the soil and

aquifer. It is therefore more meaningful to determine the major geochemical reactions that likely caused fractionation processes instead of applying a normalization correction. Groundwater ^{14}C data can be converted from pM to pmC using the measured δ^{13}C value and the following equations (Stuiver and Pollach, 1977; Mook and van der Plicht, 1999; Plummer and others, 2004):

$$pmC = \left(\frac{\delta^{14}C}{1000} + 1 \right) 100 \qquad (18)$$

$$\delta^{14}C = \left[\frac{\left(1 + \frac{\Delta^{14}C}{1000} \right) \left(1 + \frac{\delta^{13}C}{1000} \right)^2}{0.975^2} - 1 \right] 1000 \qquad (19)$$

For convenience, TracerLPM includes a workbook function called 'ConvertC14pMtoC14pmC' that can be used to convert ^{14}C reported in pM to pmC.

The carbon component used for modeling of ^{14}C in TracerLPM is not specified, but typically it will be dissolved inorganic carbon (DIC). Therefore, the ^{14}C concentration of mixtures is dependent on the concentration of ^{14}C and DIC of every parcel of water contained in that mixture. All LPMs, except the binary mixing models, assume the concentration of DIC was the same for all parcels of water in the sample, which could be invalid in some cases. Those situations would be modeled better by using a BMM. For BMMs, users must specify estimated concentrations of DIC in each of the two mixing components so that the ^{14}C value of the overall mixture is correctly calculated:

$$^{14}C = \left(f_1 \, ^{14}C_1 DIC_1 + \left(1 - f_1 \right) \, ^{14}C_2 DIC_2 \right) / DIC_{mixture} \qquad (20)$$

where

$^{14}C_1$ is carbon-14 concentration of first component,

$^{14}C_2$ is carbon-14 concentration of second component,

DIC_1 is dissolved inorganic carbon concentration of first component,

DIC_2 is dissolved inorganic carbon concentration of second component, and

$DIC_{mixture}$ is dissolved inorganic carbon concentration of mixture or DIC1 + DIC2.

Entering the same value of DIC for the young and old fractions removes the dependence of ^{14}C on DIC (see equation 15).

Helium-4

Helium-4 is modeled assuming radiogenic [4]He accumulation in groundwater occurs gradually, as a linear function of travel time. Consequently, only the radiogenic portion of measured helium provides a meaningful comparison to the LPM outlet concentrations. This requires users to enter [4]He concentrations that have been adjusted to remove contributions from equilibrium with the atmosphere, excess air, crustal fluxes, and mantle helium. Then, [4]He can be used as a tracer of groundwater age, given information about the [4]He accumulation rate in the aquifer. These adjustments still might not give plausible [4]He concentrations in groundwater because of the difficulty in correctly quantifying the many sources of helium mentioned above. In those cases, TracerLPM can be used to evaluate assumptions about sources and accumulation rates of [4]He.

One option for estimating the [4]He accumulation rate is to assume [4]He is released to pore fluid at the same rate it is produced by radioactive decay of uranium (U) and thorium (Th) in the aquifer solids. This rate can be estimated with information about the U and Th concentrations, porosity, and sediment density of the aquifer. Summaries of U and Th concentrations in surficial materials (rock and sediment) of the conterminous United States are provided in the workbook as a guide to estimate concentrations of U and Th in aquifers that are composed of similar rock types or sediment. These data were generated for TracerLPM by combining aerial gamma-radiation data from Phillips and others (1993) and geographic information system (GIS) features of the generalized geologic map of the United States (Reed and Bush, 2005). Site-specific U and Th concentrations in aquifer solids can be estimated by using local GIS features and the U and Th data set of Phillips and others (1993). Measurements on core samples are the best source for concentrations of U and Th in aquifer solids.

If [4]He produced by radioactive decay is released at the same rate to the pore fluid, then the [4]He solution rate in a groundwater parcel is determined by the following equation (Andrews and Lee, 1979), which uses the U, Th, porosity, and bulk density values provided by the user:

$$^4\text{He}(\Delta t) = \frac{\Delta t \rho}{\varphi}\left[1.19^{-13}\left(\text{U}\right) + 2.88^{-14}\left(\text{Th}\right)\right] \tag{21}$$

where

$^4\text{He}(\Delta t)$ is the helium-4 concentration in water,

     in $\dfrac{\text{cc@STP}}{\text{g of H}_2\text{O}}$,

Δt is age increment, in years,

ρ is bulk density, in grams per cubic centimeter,

φ is porosity, cubic centimeter per cubic centimeter,

U is uranium concentraton in parts per million (ppm), and

Th is thorium concentration in ppm.

Alternatively, the user can specify a [4]He solution rate (in cubic centimeters at standard temperature and pressure per gram of water per year) in TracerLPM. This will cause the program to use the user-defined [4]He solution rate instead of the [4]He production rate defined by U and Th concentrations. This option could be necessary where the rate of [4]He released to pore fluids is either faster or slower than the production rate because of non-steady-state processes related to geologic disturbance and geochemical reactions (Andrews and others, 1982, Torgensen and Clarke, 1985; Solomon and others, 1996). In some cases, the accumulation rate of [4]He in pore fluid can be determined from experimental data or estimated by comparison with other age tracers (Solomon and others, 1996). For example, apparent ^{14}C ages could be used to estimate the local [4]He accumulation rate over the time scale of ^{14}C dating, and that rate could be assumed to be valid over a longer time scale where ^{14}C dating is not possible.

Total Mean Age, Mean Age, and Unsaturated-Zone Travel Time

In this program, the total mean age of a sample is defined as the mean age of groundwater in the saturated zone plus the travel time through the unsaturated zone (UZ). As this implies, when the travel time through the UZ is zero or less than a year, the total mean age is roughly equal to the mean age of the saturated zone travel times. For BMMs, the total mean age is given by the following:

$$\tau_{total} = f_1 \tau_1 + (1 - f_1) \tau_2 + UZ_{tt} \qquad (22)$$

where

τ_{total} is total mean age of mixture,
f_1 is fraction of first component,
τ_1 is mean age of first component,
τ_2 is mean age of second component, and
UZ_{tt} is unsaturated-zone travel time.

Unsaturated-zone transport of tracers is modeled by assuming piston-flow behavior. As such, a UZ travel time of 5 years causes ^3H to enter the saturated zone 5 years after it entered the ground as precipitation. Because ^3H is a radioactive tracer, the concentration that enters the aquifer can be significantly lower than the initial ^3H concentration in precipitation. In addition, because $^3He_{trit}$ typically is lost to the atmosphere until the water parcel crosses the water table and is isolated from atmospheric exchange, long travel times of UZ transport of ^3H result in lower peak $^3He_{trit}$ concentrations in the aquifer.

It is important to note that dispersion can affect the transport of ^3H and other tracers in the unsaturated zone. Although dispersion in the UZ is not explicitly programmed as an option in TracerLPM, it is possible to create a tracer input function that accounts for dispersion by applying a dispersion model to the original tracer input history and entering the model results into the *TracerInput* worksheet.

TracerLPM allows users to set the UZ travel time for each tracer independently to a constant value specified on the *TracerInput* worksheet. This gives the user the flexibility to define different UZ travel times for tracers that appear to have age discordance in groundwater that cannot be accounted for by changes in LPM model parameters. For example, ^3H and other soluble tracers can travel through the UZ as aqueous species with infiltrating water, while relatively insoluble atmospheric gas tracers, such as CFCs, can travel more rapidly to the water table in UZ air. Consequently, it can be necessary to apply a different UZ travel time to the aqueous tracers compared to the atmospheric gas tracers in order to model ^3H and CFC concentrations in groundwater correctly.

TracerLPM also allows users to set the UZ travel time for one or more tracers to a UZ travel-time parameter that is defined on several other worksheets. This saves the user from having to change the UZ travel time for each of the tracers individually on the *TracerInput* worksheet while testing scenarios from the output and graph worksheets. The treatment of UZ travel time for each tracer can be set to a constant value specified on the *TracerInput* worksheet or set to be controlled by the UZ travel-time parameter on other worksheets.

TracerLPM Workbook

TracerLPM is an interactive Excel workbook (for Excel versions 2007 or later) program that is intended to be used to determine the LPM that most accurately describes or conceptualizes tracer concentrations measured in groundwater. The overall approach of TracerLPM emphasizes conceptualization and evaluation of the models through visual inspection of modeled and measured tracer concentrations. The workbook includes graphical routines for comparing data and calculations in multiple formats, such as Tracer-Tracer plots and Time-Series plots. Prior to modeling, these plots can be used to assess tracer data for anomalies related to local contamination, degradation, or potential artifacts of sampling or analysis.

An LPM most often is determined by selecting a model based on the conceptualization of the physical system, and inversely fitting the output concentrations to measured concentrations by varying the mean age and other model parameters. This is accomplished by graphically viewing the output tracer concentrations against modeled tracer concentrations or by subroutines that optimize a specified LPM to the measured tracer concentrations. Defining the LPM, mean age, and other model parameters provides a plausible estimation of the age distribution of a sample. The age distribution then can be viewed and used to forecast or evaluate the vulnerability of the sample site or spring to contamination from groundwater.

The TracerLPM workbook was developed by using Microsoft Visual Basic® for Applications (VBA), and the Excel® add-in that is distributed with the workbook was written in the C++ programming language by using Microsoft Visual Studio 2010 (version 10) and Microsoft Office Excel 2010 XLL Software Developer Kit. The workbook was developed for Excel® 2007 versions or later (see the "Compatibility" section in appendix B for more details).

The VBA code in this workbook primarily was developed to help automate the routine tasks in building LPM formulas and for viewing the output concentrations. Much of the code, however, relies on the stability of the current configuration of each worksheet. Therefore, changes to the worksheets, such as insertion or deletion of rows, columns, or cells, can have undesirable effects and render the workbook program inoperable. To help users avoid this event, cells have been color coded on each worksheet. In general, cells that are colored grey should not be altered, especially if the cell contains a formula. Cells colored light blue display output for certain programs and should not be altered. Cells that are colored white, or do not have a fill color attributed to the cell, can be changed. White or no-fill cells are normally used for entering data or changing model parameters.

Users can add new worksheets and graphs as needed. There are graphs on several existing worksheets, however, that are populated by VBA routines. Removing the graphs distributed with the workbook should not be done. The graphs can be altered, but that should be done with caution. Adding or removing plotted data could cause the program to not operate properly. It is recommended that changes to the graphs be confined to axis formats, axis titles, and graph layouts.

All of the graphs in this workbook have minimum and maximum values for each axis that can be changed by entering the minimum and maximum values in corresponding cells to the right of or below each graph. In addition, the axes can be scaled automatically by clicking the check box next to each graph. If the auto-scale check box is left unchecked, then subsequent changes to the worksheet that cause changes to the graphs, such as selection of additional well to graph, will not affect the minimum and maximum values of the axes.

Determining the Lumped Parameter Model

The general procedure for determining the LPM and mean age of a sample is to select a model based on the conceptualization of the physical system, and vary the mean age and any additional model parameters until the model output concentration matches the measured concentrations. The determination of an LPM from a single tracer, such as tritium or CFCs, measured in a single sample is non-unique. The reason for this is both tritium and CFCs have input histories whose concentrations increase, peak, and decrease over time. As a result, measured tritium or CFC concentrations can correspond to more than one mean age. In these cases, and in general, the best estimation of the age distribution and mean age is made by using multiple tracers analyzed in a single sample (Tracer-Tracer method) or from a single tracer analyzed in multiple samples collected over a period of time (Time-Series method). TracerLPM permits optimization of parameters in either of these modes, including multiple tracers in multiple samples over time.

Users are encouraged to choose models that are most similar to the hydrogeologic situation (see figs. 1 through 6). If the user is unsure about which model to use, it is recommended to begin by plotting the simplest models (PFM and EMM) with the measured sample concentrations. The PFM and EMM have no other parameters besides mean age; therefore, sample tracer concentrations should plot near the PFM or EMM line if either model is an acceptable choice for a sample. The PFM assumes no mixing within the aquifer or well, while the EMM assumes complete mixing in the well or spring. Sample concentrations that fall off the PFM and EMM curves can indicate other age distribution models (e.g., EPM, PEM, or DM) could be more appropriate. The EPM, PEM, and DM have additional parameters (EPM ratio, PEM ratio, and the dispersion parameter, DP) that significantly affect the output concentrations of each model and should be adjusted for each sample. EPM and PEM ratios greater than 1 and DP values less than 0.1 tend to produce more PFM-like behavior, whereas EPM and PEM ratios less than 0.1 and DP values greater than 1 produce more EMM-like behavior. Samples with tracer concentrations that do not plot near any of the previous four LPMs even after parameter adjustments could indicate the sample is a binary mixture. Binary mixtures can be evaluated by specifying ages for each component and varying any other model parameters. For most tracers, the output from the BMMs will produce a straight line originating from the output concentration of the second component and ending at the first component.

It is important to emphasize that tracer data for some samples will not plot near any of the model curves, either because the age distribution in the sample is not like any of the models or because of various problems associated with the tracers. Gross discrepancies between data and models could indicate tracer contamination or degradation in the aquifer, degassing, artifacts of sampling and handling, and analytical errors, among other things (International Atomic Energy Agency, 2006). Users should be aware of these potential problems, and how they can be recognized with the aid of Tracer-Tracer plots and other geochemical evidence. Problematic data can be eliminated from consideration in the TracerLPM fitting routines, but this should be done with caution and with appropriate documentation.

Once a model has been identified and its parameters estimated (by visual inspection or independent knowledge about the site), the best-fit algorithms on the *TracerTracerFits* and *TimeSeriesFits* worksheets can be used to calculate the mean age of the sample. Because most of the tracers have transient and non-linear recharge histories, multiple local minima in the residual errors for the best-fit models can exist. This difficulty often makes estimating groundwater age by using the best-fit tool impractical without some prior knowledge of the conceptual appropriateness of the selected LPM and an initial estimate for the mean age and any other model parameters that provide a relatively close match between modeled and measured sample concentrations. For this reason, it is recommended that users select an LPM on the basis of a conceptual understanding of the flow system, and graphically estimate the model parameter values for each sample prior to using the best-fit tool to refine the age estimate.

The best-fit models are evaluated by calculating the total difference between the modeled and observed tracer concentrations. The total difference is measured either by relative error or relative squared error. The acceptance criteria for choosing to keep or reject a model solution depends on the choice of measurement error type, the number of tracers used in the analysis, and the number of parameters. In general, the relative error is a more restrictive and accurate measure of the difference between modeled and observed tracer values than the relative squared difference. The relative squared error is frequently used in fitting procedures, however, and is therefore provided for comparison purposes.

Ideally, the number of tracer measurements used to fit the model should be at least "$p + 1$," where "p" is the number of parameters in the LPM. For the PFM and EMM, p is 1. For the DM, EPM, and PEM, p is 2. For BMMs, p is equal to the number of parameters in the first model, plus the number of free parameters in the second model, plus 1 for the mixing fraction. So in the simplest case, such as a BMM-PFM-EMM or a BMM-PFM-PFM, p would be 3. Likewise, p would be 5 for the most complicated case, such as a BMM-DM-DM or a BMM-DM-PEM. Commonly, however, it is not feasible to collect six tracers at a well or spring that is characterized by a BMM. In those cases, it can be necessary to use tracer data at other wells in the same aquifer to characterize and constrain the model parameters of one of the mixtures in the BMM (for example, Bexfield and others, 2012). In other cases, it can be difficult to constrain a model well, so users should be cautious in asserting the validity of a particular model.

SetupWorkbook Worksheet

Opening the workbook causes the program to automatically display the *SetupWorkbook* worksheet (fig. 8). The *SetupWorkbook* worksheet is the starting point for

configuring the workbook so that worksheets necessary for completing certain tasks are visible, and worksheets that are not used are hidden from view. There are three workbook configurations: Tracer-Tracer, Time-Series, Age Distribution & Forecasting. The Tracer-Tracer workgroup is used for determining the LPM and mean age of a sample by using data for one or more tracer collected on a single sampling date. The Time-Series workgroup is used to determine the mean age and the age distribution for a spring or well by using a single (or multiple) tracers collected on multiple sample dates. The Age Distribution & Forecasting workgroup is used for viewing user-defined age distributions or LPM age distributions, and for understanding historic, and forecasting future, water-quality changes at a site. The worksheets that make up each workgroup are organized so that the user works through each worksheet from left to right beginning with the *Samples* worksheet. The worksheet tabs have been color coded according to the workgroup to which it belongs. The *SetupWorkbook*, *Samples*, *TracerInput*, and *SavedModelAges* worksheets are common to all configurations and are colored green.

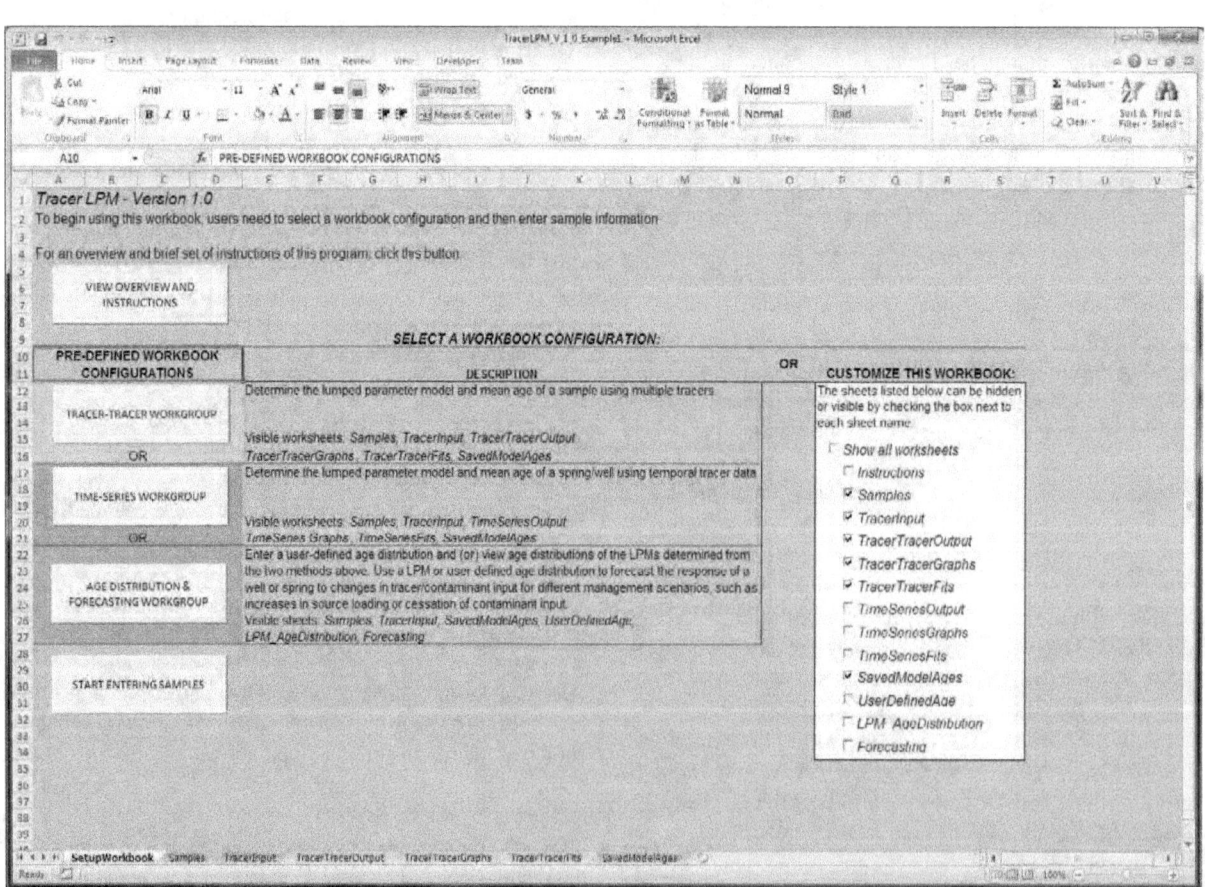

Figure 8.   Screenshot of the *SetupWorkbook* worksheet.

Samples Worksheet

The *Samples* worksheet is used to select tracers to model and enter sample information and tracer concentrations (fig. 9). It is expected that tracer concentrations entered into the workbook have been corrected for recharge temperature, recharge elevation, excess air, equilibrium with the atmosphere, terrigenic helium, or degradation, when applicable.

Tracers for which sample data are available are selected from a set of 10 pull-down menus (fig. 9). The tracers listed in each pull-down menu are the most frequently collected by the USGS; however, additional tracers can be added to the workbook (see *Tracers* Worksheet section). When a tracer is selected that has stored tracer data, a dialog box asks the user to select the location for the tracer input history (recharge concentrations) that correspond to the samples of interest. If the relevant tracer input concentrations already exist in the *StoredTracerData* worksheet, the user can select the desired location (fig. 10).

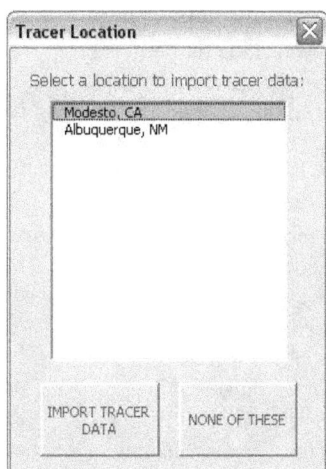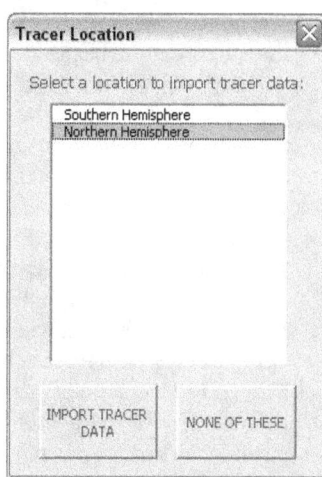

Figure 10. Screenshot of the dialog boxes for selecting tracer input history locations.

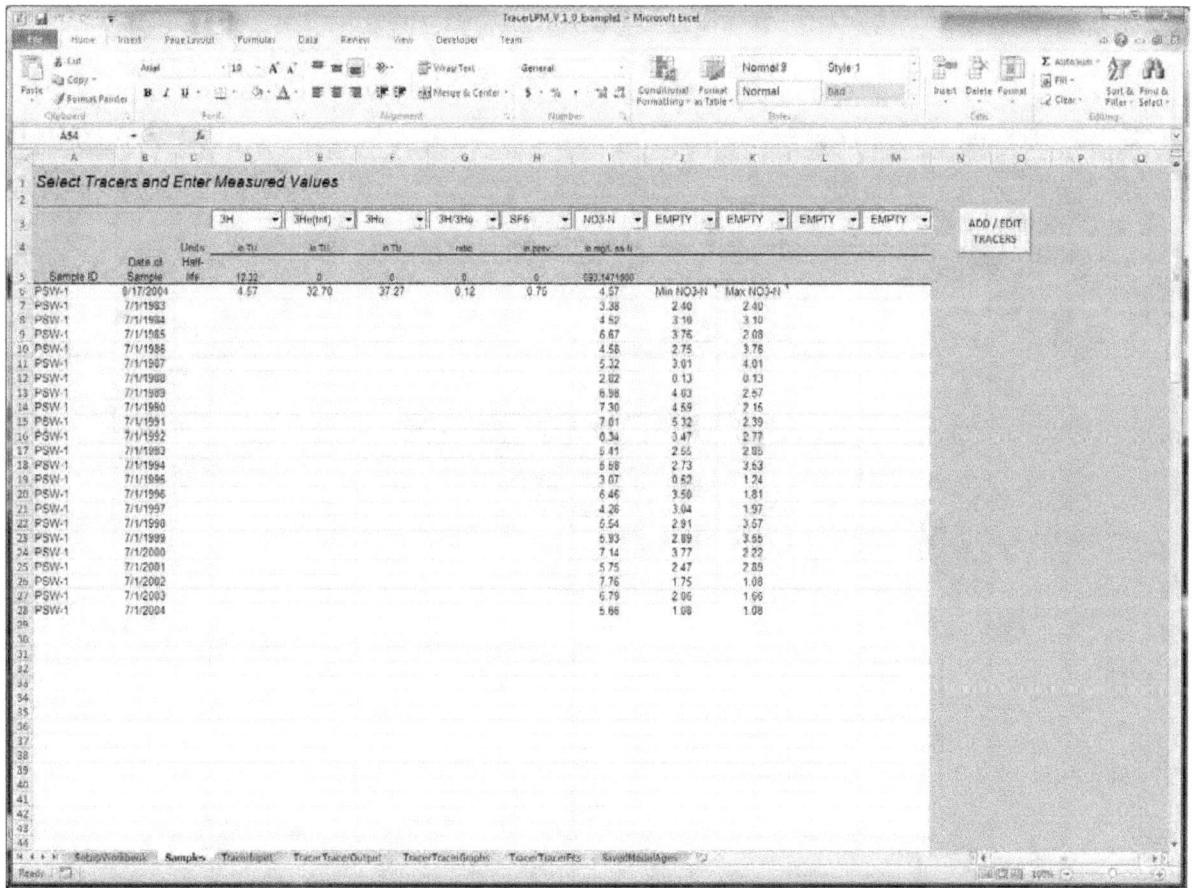

Figure 9. Screenshot of the *Samples* worksheet.

The *StoredTracerData* worksheet is normally hidden from view but can be made visible by clicking the "VIEW STROED TRACER DATA" button in the upper left-hand corner of the *TracerInput* worksheet. If the input history has not been pre-loaded, the user is asked to load the tracer data into the '*TracerInput*' worksheet (see *Tracers* and *TracerInput* worksheets). Actual sample data for the well or spring are manually entered into the *Samples* worksheet in the cells below the corresponding tracer heading.

It is advantageous only to select tracers from the *Samples* worksheet that will be modeled, and select the "EMPTY" tracer in those columns that will not be used (fig. 9). The inclusion of an unused tracer in any of the columns D through M will force the calculation of that tracer in the *TracerTracerOutput* or the *TimeSeriesOutput* worksheet. The speed at which Excel can calculate the model output in the workbook is directly dependent on the number of cells that require calculation. Hence, reducing the number of modeled tracers (where appropriate) also reduces the number of calculations and decreases the time between successive calculations.

Tracers Worksheet

The tracers listed in the pull-down menus and their half-life and decay constants are stored in a hidden worksheet called *Tracers* (fig. 11). To unhide and access the *Tracers* worksheet, click the "ADD / EDIT TRACERS" button on the *Samples* worksheet (fig 9). The information in the *Tracers* worksheet does not change frequently, so the worksheet is hidden to minimize the number of sheets displayed in the workbook.

The half-life and decay constants are declared globally from the *Tracers* worksheet. The user can add new tracers or edit existing information associated with any previously loaded tracers. The actual tracer input history for a newly added tracer is entered elsewhere (see TracerInput Worksheet).

The *Tracers* worksheet also defines how the unsaturated travel time is modeled for each tracer. Entering "Constant" causes the program to link the unsaturated travel-time parameter in an LPM to the unsaturated travel-time value in row 9 on the *TracerInput* worksheet. Entering "Vary by UZ parm" causes the program to link the unsaturated travel-time

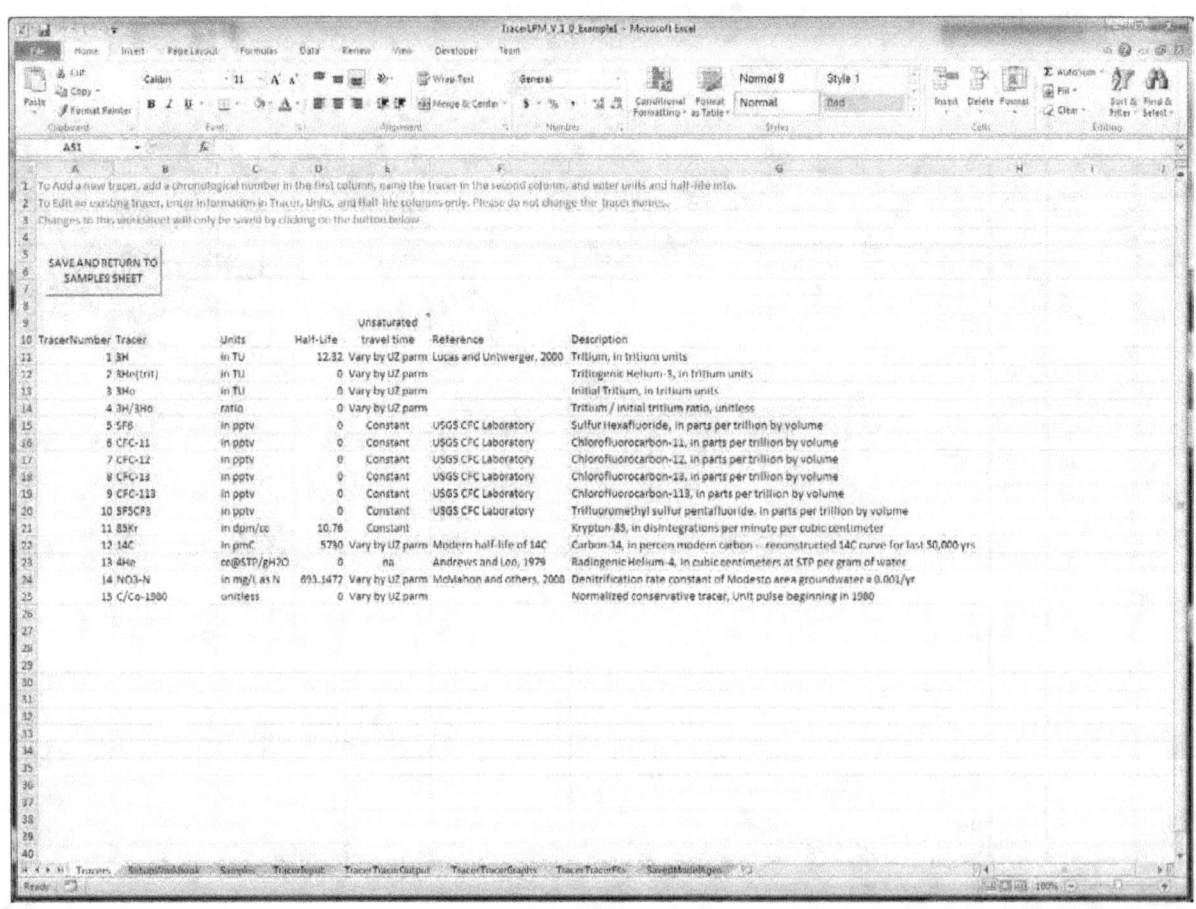

Figure 11. Screenshot of the *Tracers* worksheet. This worksheet is normally hidden from view.

parameter in an LPM to the unsaturated travel-time parameter cell located on various other worksheets. This allows users to fix some tracers to one UZ value and vary the UZ time for other tracers.

When finished entering information, click the button, "SAVE AND RETURN TO SAMPLES SHEET," in the upper left-hand corner of the *Tracers* worksheet to return to the *Samples* worksheet. New tracers will now be included in the tracer pull-down menus, and any changes made to the half-life or units of the existing tracers will be reflected in rows 4 and 5 and other places throughout the workbook if the tracer was already selected. The workbook can store the input history of tracers added to the workbook by going to the *TracerInput* worksheet and following the instructions for adding tracer input histories under the TracerInput section of this manual.

TracerInput Worksheet

The *TracerInput* worksheet is used for viewing and entering the tracer input history in recharge for each tracer selected from the *Samples* worksheet (fig. 12). The

concentrations of dissolved inorganic carbon (DIC) needed for the ^{14}C mixing calculation, and input parameters for determining the ^4He accumulation rate, are defined to the right of the tracer input history.

For tracers that have been defined to have a constant UZ travel time, users can enter the constant value in row 9 of the *TracerInput* worksheet. Tracers that are specified as "Vary by UZ parm" are controlled by the value entered into the UZ travel-time parameter cell defined on other worksheets.

When a tracer has been permanently saved and stored to the workbook, the tracer data can be scaled relative to the stored values by changing the value in row 17 on the *TracerInput* worksheet. In the case of ^{14}C, decreasing the scaling factor allows users to simulate the dilution of ^{14}C from geochemical reactions occurring in the unsaturated or saturated zone. For other tracers, the scaling factor can be used to simulate local variations of tracer input concentrations in recharge, which can be either enriched or diluted relative to the tracer input history.

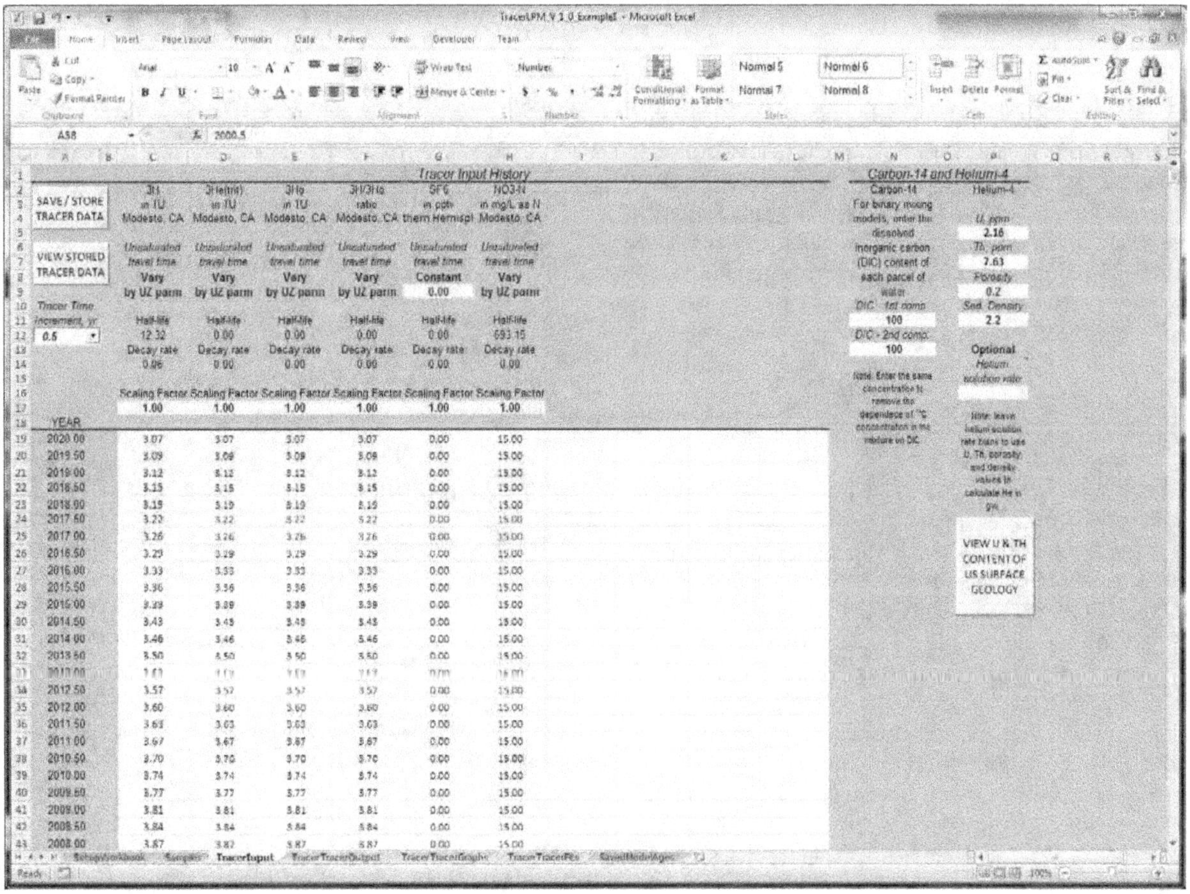

Figure 12. Screenshot of the *TracerInput* worksheet.

Adding Tracer Input Data

The input history for CFC-11, CFC-12, CFC-13, CFC-113, SF_6, SF_5CF_3, and ^{14}C were pre-loaded into this workbook. The input history of tracer concentrations in recharge for these tracers will be automatically loaded into the *TracerInput* worksheet upon selection from the *Samples* worksheet. Data for the tritiogenic tracers or other tracers that are not pre-loaded in the TracerLPM workbook must be entered manually in columns C through L of the *TracerInput* worksheet. Before entering tracer input histories into these columns, an entry for a new tracer must exist on the *Tracers* worksheet, which is accessed through the *Samples* worksheet. The header information for any new tracer in rows 2–3 and 6–14 of the *TracerInput* worksheet (fig. 12) is automatically populated from the *Tracers* worksheet when the tracer is selected from one of the pull-down menus on the *Samples* worksheet (fig. 9).

For this workbook, integration of the convolution equation (1) is performed on a monthly basis. Therefore the input history entered into the workbook should be the average value for the month, if monthly data are available. This distributes the error of integration evenly before and after the mid-point of the month. Consequently, users should be aware that tracer input data can need to be offset before it is entered into this workbook.

To enter concentrations in recharge for a newly added tracer, the time increment should be changed to conform to the new tracer input data. The user then enters the tracer input data into the corresponding time span. In most instances, tracer concentrations in recharge will not be available for the entire period of record (1850 to 2020). Estimated values must be entered for missing periods. For periods beginning in 1850 and prior to the start of the new tracer input history, entered values should be estimated as the "pre-tracer" or native tracer concentration. For the period beginning after the end of the new tracer input history and 2020, estimated values are often defined as either constant, increasing, or decreasing from the last measured value entered.

For the tritiogenic tracers (^3H, $^3He_{trit}$, ^3H_o, and $^3H/^3H_o$), users only need enter the input history for tritium. Users should select only tritium from the *Samples* worksheet and enter the tritium input data for the location of interest on the *TracerInput* worksheet. Once tritium data are stored to the workbook, the other tritiogenic tracers ($^3He_{trit}$, ^3H_o, and $^3H/^3H_o$) can be selected from the *Samples* worksheet. The other tritiogenic tracers use a copy of the tritium data. The program internally calculates the accumulation of $^3He_{trit}$ and the $^3H/^3H_o$ ratio from the tritium input data. The only difference between ^3H_o and ^3H is that the decay constant for ^3H_o is zero. For binary mixtures involving ^{14}C, the user will need to specify the concentration of DIC in the two mixtures in cells N11 and N13.

Helium-4 concentrations in groundwater are calculated from the uranium, thorium, porosity, and sediment density values specified in cells P5, P7, P9, and P11 by using equation 21. Alternatively, users can enter a helium solution rate (in cubic centimeters at standard temperature and pressure per gram of water per year) in cell P16. Leave this cell blank to use the calculated helium solution rate based on uranium and thorium values. Summaries of uranium and thorium in rock types of the U.S. can be viewed by clicking the button in column P.

Storing Tracer Input Data

The input history of tracers is stored in the hidden worksheet called *StoredTracerData* (fig. 13). This worksheet is normally hidden from view but can be made visible by clicking the "VIEW STORED TRACER DATA" button in the upper left-hand corner of the *TracerInput* worksheet. Generally, this worksheet should not be manually altered by the user. The program has routines for altering and storing tracer data to this worksheet. To store new or altered input histories of tracers in recharge, a location must be entered in row 4 of the tracer column on the *TracerInput* worksheet. The user then clicks the button, "SAVE / STORE TRACER DATA," in the upper left-hand corner of column A. For each tracer listed in columns C through L, the program will check to see if tracer data at that location are already stored in the workbook. If tracer data do not exist at that location, the program will write the tracer data in monthly increments to the *StoredTracerData* worksheet to store the data. If tracer data are already stored in the workbook, the program will check the overwrite value in row 3 of the *StoredTracerData* worksheet. If the value is not set to "Locked," the user will be asked if the program should overwrite those data. Otherwise, the program will not save the tracer data for "Locked" data. This prevents the accidental overwrite of the tracer data that were distributed with the program.

Time Increments

A pull-down menu in the upper-left hand corner of the *TracerInput* worksheet controls the time increment of the data listed in the white data cells (fig. 12). The user has the choice of monthly (0.08333), quarterly (0.25), semi-annual (0.5) and yearly (1) increments. The time increment will be applied to a starting date of 1850 and end in year 2020. Each time the time increment is changed, the program will automatically reload all tracer data stored in the *StoredTracerData* worksheet into the *TracerInput* worksheet to conform to the new time increment. If the time increment is changed to a more coarsely discretized increment than the one used in compiling the input data, the workbook will calculate the

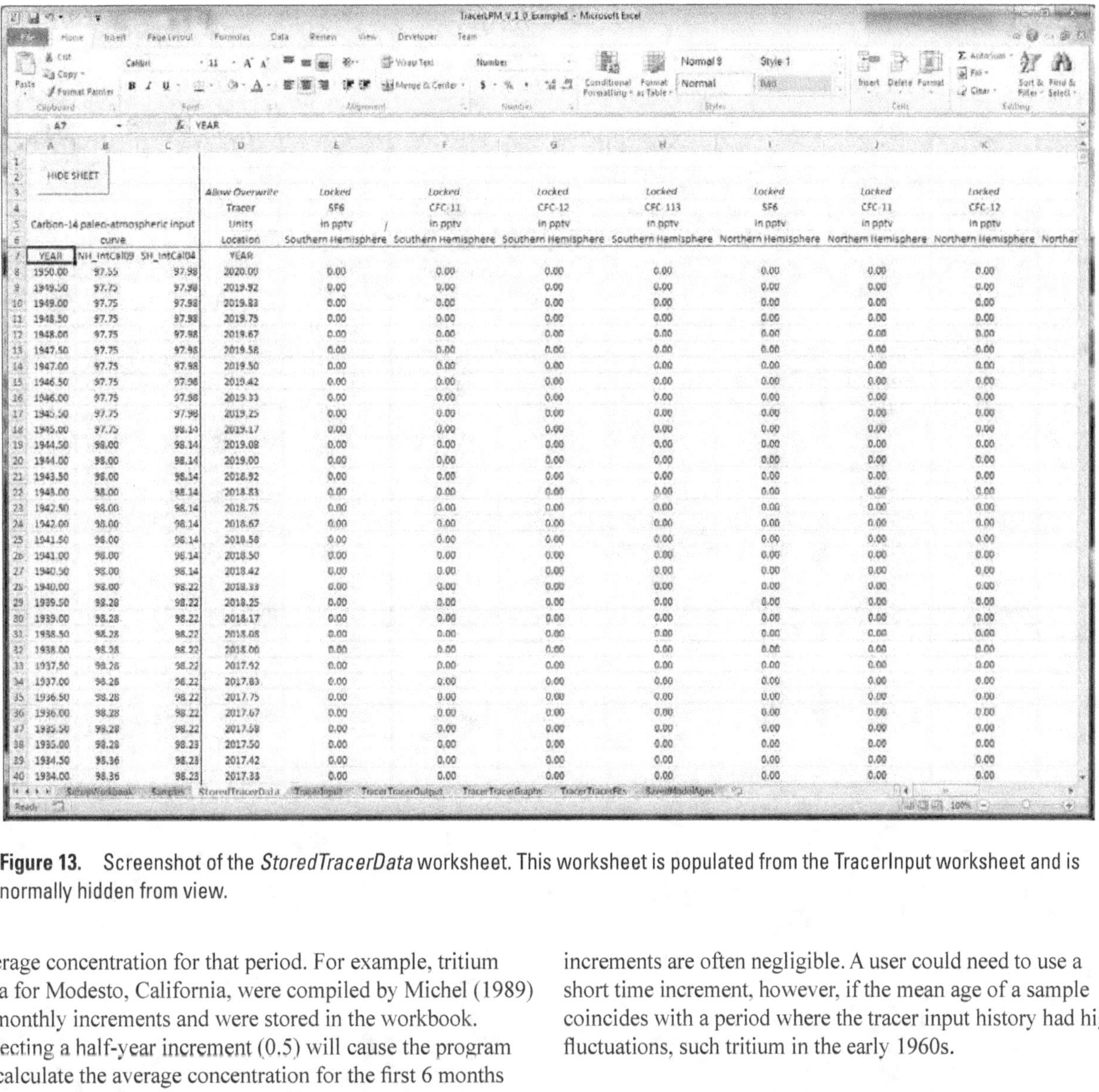

			Allow Overwrite	Locked	Locked	Locked	Locked	Locked	Locked	Locked
			Tracer	SF6	CFC-11	CFC-12	CFC-113	SF6	CFC-11	CFC-12
Carbon-14 paleo-atmospheric input curve			Units	in pptv	in pptv	in pptv	in pptv	in pptv	in pptv	in pptv
			Location	Southern Hemisphere	Southern Hemisphere	Southern Hemisphere	Southern Hemisphere	Northern Hemisphere	Northern Hemisphere	Northern Hemisphere
YEAR	NH_IntCal09	SH_IntCal04	YEAR							
1950.00	97.55	97.98	2020.00	0.00	0.00	0.00	0.00	0.00	0.00	0.00
1949.50	97.75	97.98	2019.92	0.00	0.00	0.00	0.00	0.00	0.00	0.00
1949.00	97.75	97.98	2019.83	0.00	0.00	0.00	0.00	0.00	0.00	0.00
1948.50	97.75	97.98	2019.75	0.00	0.00	0.00	0.00	0.00	0.00	0.00
1948.00	97.75	97.98	2019.67	0.00	0.00	0.00	0.00	0.00	0.00	0.00
1947.50	97.75	97.98	2019.58	0.00	0.00	0.00	0.00	0.00	0.00	0.00
1947.00	97.75	97.98	2019.50	0.00	0.00	0.00	0.00	0.00	0.00	0.00
1946.50	97.75	97.98	2019.42	0.00	0.00	0.00	0.00	0.00	0.00	0.00
1946.00	97.75	97.98	2019.33	0.00	0.00	0.00	0.00	0.00	0.00	0.00
1945.50	97.75	97.98	2019.25	0.00	0.00	0.00	0.00	0.00	0.00	0.00
1945.00	97.75	98.14	2019.17	0.00	0.00	0.00	0.00	0.00	0.00	0.00
1944.50	98.00	98.14	2019.08	0.00	0.00	0.00	0.00	0.00	0.00	0.00
1944.00	98.00	98.14	2019.00	0.00	0.00	0.00	0.00	0.00	0.00	0.00
1943.50	98.00	98.14	2018.92	0.00	0.00	0.00	0.00	0.00	0.00	0.00
1943.00	98.00	98.14	2018.83	0.00	0.00	0.00	0.00	0.00	0.00	0.00
1942.50	98.00	98.14	2018.75	0.00	0.00	0.00	0.00	0.00	0.00	0.00
1942.00	98.00	98.14	2018.67	0.00	0.00	0.00	0.00	0.00	0.00	0.00
1941.50	98.00	98.14	2018.58	0.00	0.00	0.00	0.00	0.00	0.00	0.00
1941.00	98.00	98.14	2018.50	0.00	0.00	0.00	0.00	0.00	0.00	0.00
1940.50	98.00	98.14	2018.42	0.00	0.00	0.00	0.00	0.00	0.00	0.00
1940.00	98.00	98.22	2018.33	0.00	0.00	0.00	0.00	0.00	0.00	0.00
1939.50	98.28	98.22	2018.25	0.00	0.00	0.00	0.00	0.00	0.00	0.00
1939.00	98.28	98.22	2018.17	0.00	0.00	0.00	0.00	0.00	0.00	0.00
1938.50	98.28	98.22	2018.08	0.00	0.00	0.00	0.00	0.00	0.00	0.00
1938.00	98.28	98.22	2018.00	0.00	0.00	0.00	0.00	0.00	0.00	0.00
1937.50	98.28	98.22	2017.92	0.00	0.00	0.00	0.00	0.00	0.00	0.00
1937.00	98.28	98.22	2017.83	0.00	0.00	0.00	0.00	0.00	0.00	0.00
1936.50	98.28	98.22	2017.75	0.00	0.00	0.00	0.00	0.00	0.00	0.00
1936.00	98.28	98.22	2017.67	0.00	0.00	0.00	0.00	0.00	0.00	0.00
1935.50	98.28	98.22	2017.58	0.00	0.00	0.00	0.00	0.00	0.00	0.00
1935.00	98.28	98.23	2017.50	0.00	0.00	0.00	0.00	0.00	0.00	0.00
1934.50	98.36	98.23	2017.42	0.00	0.00	0.00	0.00	0.00	0.00	0.00
1934.00	98.36	98.23	2017.33	0.00	0.00	0.00	0.00	0.00	0.00	0.00

Figure 13. Screenshot of the *StoredTracerData* worksheet. This worksheet is populated from the TracerInput worksheet and is normally hidden from view.

average concentration for that period. For example, tritium data for Modesto, California, were compiled by Michel (1989) at monthly increments and were stored in the workbook. Selecting a half-year increment (0.5) will cause the program to calculate the average concentration for the first 6 months of the year and for the second 6 months of the year and load these two values in the *TracerInput* worksheet for each half-year increment. Conversely, if tracer data are compiled on a yearly basis, and a half-year increment (0.5) is chosen, then the program assumes the tracer input was constant over the first and second 6 months of each year. For calculation speed, it is advantageous to use longer time increments (0.5, 1) than shorter ones (less than 0.5). Although there is a loss in precision associated with longer time increments, differences in computed mean ages from the different time increments are often negligible. A user could need to use a short time increment, however, if the mean age of a sample coincides with a period where the tracer input history had high fluctuations, such tritium in the early 1960s.

Tracer-Tracer Workgroup

The Tracer-Tracer workgroup is used for determining the mean age of a single sample that has multiple tracers measured from the sample. The Tracer-Tracer workgroup consists of three worksheets in addition to the *Samples*, *TracerInput*, and *SavedModelAges* worksheets: *TracerTracerOutput*, *TracerTracerGraphs*, and *TracerTracerFits*. The tabs of these worksheets are colored orange.

TracerTracerOutput Worksheet

The *TracerTracerOutput* worksheet can be used to select lumped parameter models that will be calibrated to the tracer data and to view the model output (fig. 14). Models are selected from the pull-down menus in column A, and model parameters are defined in columns B through G. The selection of models and changes to model parameters in this worksheet are linked to the models and model parameters listed on the *TracerTracerGraphs* worksheet in columns H through O. The sample date is a numeric, decimal fraction of the year and should not be entered as the date format of "3/06/2003." To determine the numeric date of a sample date, the following formula can be entered into cell B14: '=DecimalYear("03/26/2003").' The workbook will calculate a date of 2003.23. Changes to the sample date in cell B14 of the *TracerTracerOutput* worksheet also will change the sample date in cell E13 of the *TracerTracerGraphs* worksheet.

Upon selection of a model, tracer output concentrations will be calculated for each mean age listed in column K and each tracer listed in row 2. Up to four models can be selected, and each model has a corresponding set of calculations in columns M through AZ. Up to 1,000 mean ages can be

modeled from column K. The mean ages listed in column K can be altered to suit a more specific range of mean ages. In addition, the user can specify a series of ages to populate column K by entering a starting age, ending age, and time-step increment in cells A21, B21, and C21. The program will incrementally add the time-step increment to the starting age until the ending age is met.

If a BMM is selected from one of the pull-down menus, the model parameter for the first component is defined in column C, and the mean age of the first component, mean age of the second component, and model parameter of the second component are defined in columns E through G, respectively. BMMs are calculated for different mixing fractions in column I and can be altered to model more specific fractions in detail. The mixing fraction defines the amount of the first component in a sample. Consequently, a mixing fraction of zero contains no first component, and a mixing fraction of 1 contains no second component in the mixture. Results from the BMMs are listed in columns BA through CN. Each column from CO through CX holds the output concentrations of the second component and is the respective input for the first component of the BMMs defined in columns BA through CN.

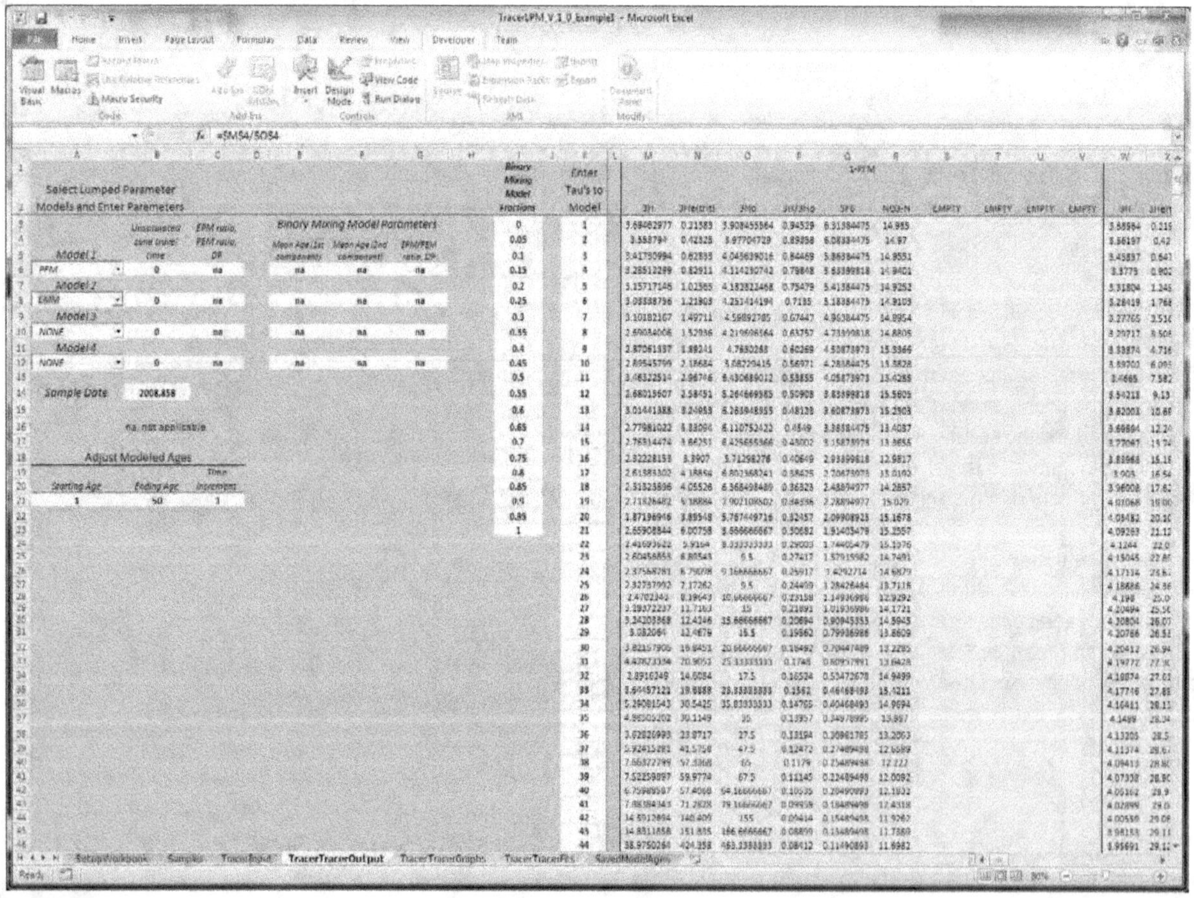

Figure 14. Screenshot of the *TracerTracerOutput* worksheet.

TracerTracerGraphs Worksheet

The *TracerTracerGraphs* worksheet is used to graphically estimate the mean age of a sample (fig. 15). The mean age of a sample is estimated by viewing tracer output concentrations for various LPMs in relation to the actual measured tracer concentrations of a sample. Users will need to select samples to graph, a sample date, and models to view. The user can create up to 10 tracer-tracer graphs to view LPM output concentrations and measured data.

TracerTracerGraphs Setup

To populate the graphs, the user selects one or more samples, a sample date, and one or more models in the upper left-hand area of the *TracerTracerGraphs* worksheet. One or multiple samples can be graphed by selecting samples from the list box in the upper-left hand corner of the worksheet. Use the 'CTRL' or 'SHIFT' key in conjunction with mouse clicks to select multiple samples. A sample date should be selected from the second list box when selecting samples to graph. The sample date is a reference time used by each LPM to calculate the output concentrations for different mean ages from that date. Because the commonly used atmospheric environmental

tracers are transient (their concentrations in recharge have changed over time), the tracer concentrations in any water sample will depend not only on the age distribution but, also, on the date the sample was collected. If more than one sample is chosen to graph, an average date will be included in the list of sample dates. If the samples were collected within a few weeks or within a couple months of each other, the average date is often an acceptable choice, if the tracer input history in recharge does not fluctuate significantly over short time scales. Because of the dramatic and substantial increase in tritium in the early 1960s, choosing an average date for samples that were collected more than 6 months apart could lead to a poor estimate of the mean age of some samples, particularly if the mean age is near the bomb-peak of the early 1960s. In those cases, it is more appropriate to model the samples by each individual date rather than by an average date.

Lumped parameter models are selected from the pull-down menus in column H. If a BMM is selected from one of the pull-down menus, the model parameter for the first component is defined in column K, and the mean age of the first component, mean age of the second component, and model parameter of the second component are defined in columns M through O, respectively.

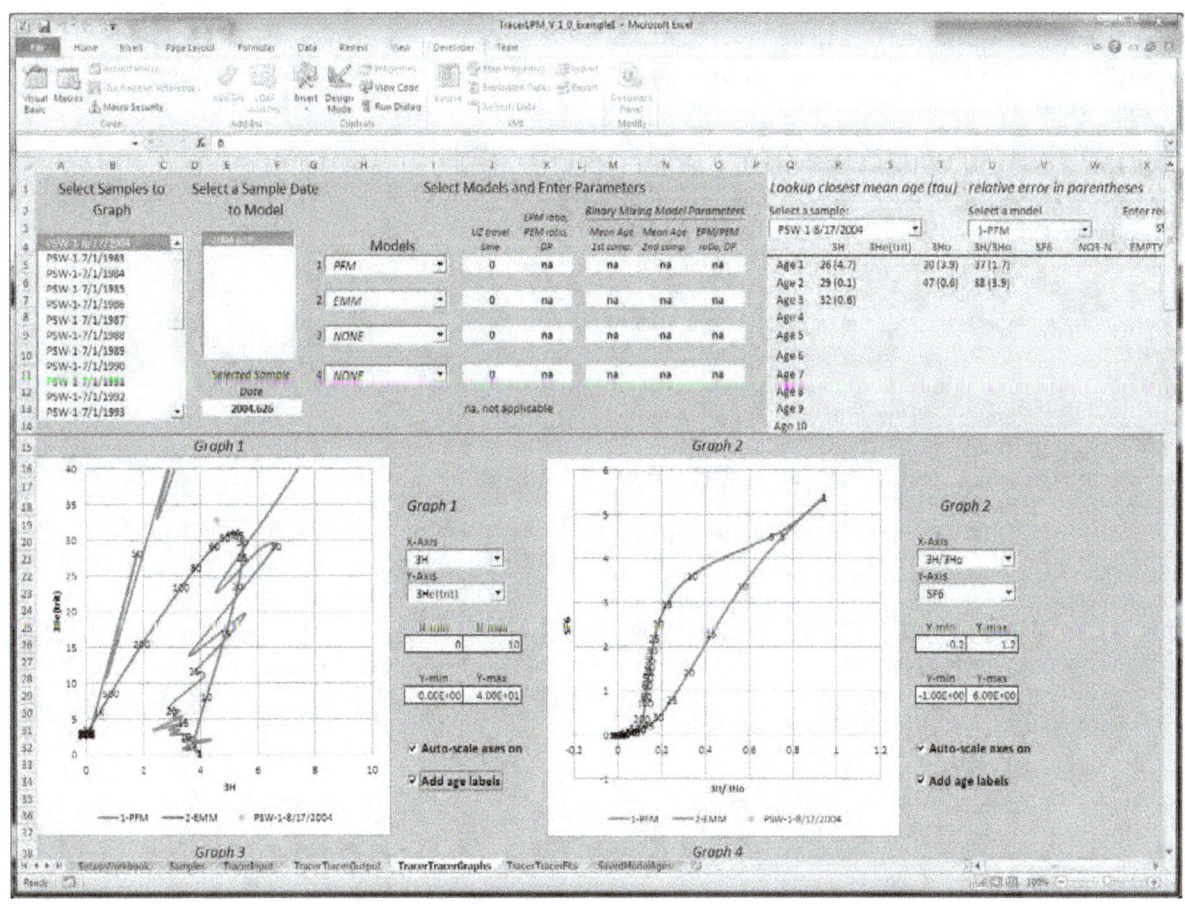

Figure 15. Screenshot of the *TracerTracerGraphs* worksheet.

Graphs

A set of 10 graphs are available for the user to create custom tracer-tracer graphs. Each graph will display a curve for each selected model (LPM) that depicts the tracer output concentrations calculated by using the model and the mean ages listed in column K of the *TracerTracerOutput* worksheet. Selected samples that have measured values for both tracers also will be plotted on the graphs.

Tracers can be selected from two pull-down menus, located to the right of each graph. The pull-down menus contain the tracers selected by the user from the *Samples* worksheet. Each graph has a legend below the x-axis with the models listed first and the samples listed second.

The mean age of the model output can be added as labels to the graphs by selecting the check box labeled "Add age labels" next to each graph (fig. 15). The labels will display the mean ages of the model output at approximately 5-year increments. For BMMs, the labels will display the fraction of the first component in the mixture. De-selecting the check box will clear the mean age and mixing fraction labels from the graphs. The mean age of the sample can be estimated visually from this procedure.

Lookup Mean Ages

The position of the model curves on the graphs change in response to changes of the model parameter values in columns J through O on the *TracerTracerGraphs* worksheet. Once a model begins to agree with measured sample data, the mean age of a sample and the relative error between the LPM output and measured data can be determined from an automated program that populates the blue-shaded area in the upper right-hand corner of the worksheet (see example that follows).

The sample and model are selected from the pull down menus in row 3 in the blue-shaded area. For BMMs, the program will output the fraction of the first component, instead of the mean age, because the mean ages for the first and second model components are already specified by the user. Relative error is the absolute value of the sample concentration minus the modeled concentration, divided by the sample concentration expressed as a percentage. The relative error for each tracer is displayed in parentheses next to each mean age. Only the first 10 mean ages found by the program are reported. If 10 mean ages are retrieved for a tracer, the cutoff value can be lowered to reduce the number of mean ages retrieved. The cutoff value for relative error can be increased if no mean ages are reported. Because some tracers have input concentrations in recharge that are not linear, multiple local minima for the mean age of a sample can exist.

As an example, figure 16 shows how the mean age was estimated for the sample in TracerLPM_V_1_Example1. xlsm, which is included in the download. Graphical evaluation of this sample found good agreement between modeled and measured concentrations by using a PEM with a ratio of 0.1. Entering a value of 5 percent yielded several mean ages (top diagram on figure 16) for the tracers ^{3}H, ^{3}He$_{trit}$, ^{3}H$_o$, and ^{3}H/^{3}H$_o$. The tracers, ^{3}H and ^{3}H/^{3}H$_o$, suggested a mean age between 60 and 70 years, while there were numerous, much younger mean ages returned for ^{3}He(trit) and ^{3}H$_o$. Although ^{3}He$_{trit}$ and ^{3}H$_o$ indicated there were several possible ages between 29 and 36, none of these ages were listed for ^{3}H or ^{3}H/^{3}H$_o$, which is an indication there is no global minimum for the sample between 29 and 36 years. Consequently, if the relative error is constrained to 3 percent (bottom diagram on figure 16), ^{3}He$_{trit}$ and ^{3}H$_o$ also have low relative errors (minimums) for ages between 60 and 70 years. This indicates the mean age of this sample is somewhere between 60 and 70 years because each tracer has a minimum or low relative error in that range of ages. From this analysis, the age can be determined more accurately by using the TracerTracerFits algorithm and the LPM and estimated mean age and model parameter. The same method can be used for determining the mixing fraction for BMMs.

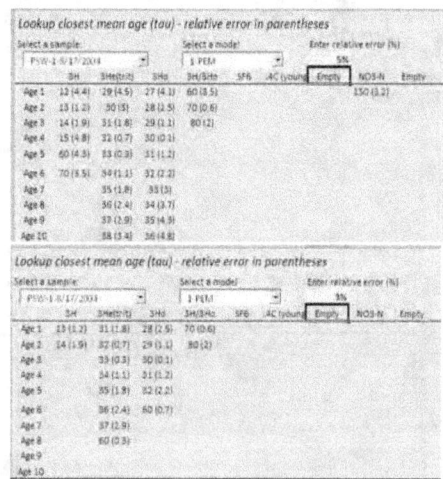

Figure 16.   Screenshot of the mean ages returned by the *TracerTacer* program for two different relative errors.

TracerTracerFits Worksheet

The *TracerTracerFits* worksheet is used to more accurately determine the mean age, model parameter, or mixing fractions for a sample (fig. 17). In some cases where the estimated age is not likely to improve by more than a year and the other parameters are not likely to deviate by more than 0.05 units, the estimated mean age and model parameter determined from the *TracerTracerGraphs* worksheet do not need refinement. The best-fit mean age, model parameter, or mixing fractions is found by minimizing the total error between LPM tracer output concentrations and measured concentrations by using a custom search algorithm and Solver (Fylstra and others, 1998).

The user will need to specify a sample, a model, and the tracers to use in the optimization. Samples are selected from the first pull-down menu in the upper left-hand part of the worksheet. The sample information and measured tracer data will be automatically populated beneath the tracer columns in row 6. The model is selected from the second pull-down menu.

The selection of a model causes the program to calculate tracer output concentrations for each tracer in row 18 using the model parameters in bold from row 12. For BMMs, the mean age and model parameter for the second model component are entered in cells N12 and O12. The output concentrations for the second model component are calculated beneath each output tracer concentration in row 19. These concentrations are used as input in the BMM calculations in row 18. This row is normally hidden from view to prevent confusion with the complete model output concentrations in row 18.

It should be noted that for BMMs, only the first model component is optimized, and the second model component is left static. It is expected that one of the model components is well enough constrained so that the other model component and mixing fraction can be optimized. It is also possible to optimize the entire BMM by optimizing one component, then reversing the BMM so that the second component is the first and optimizing that model. This method can require multiple iterations.

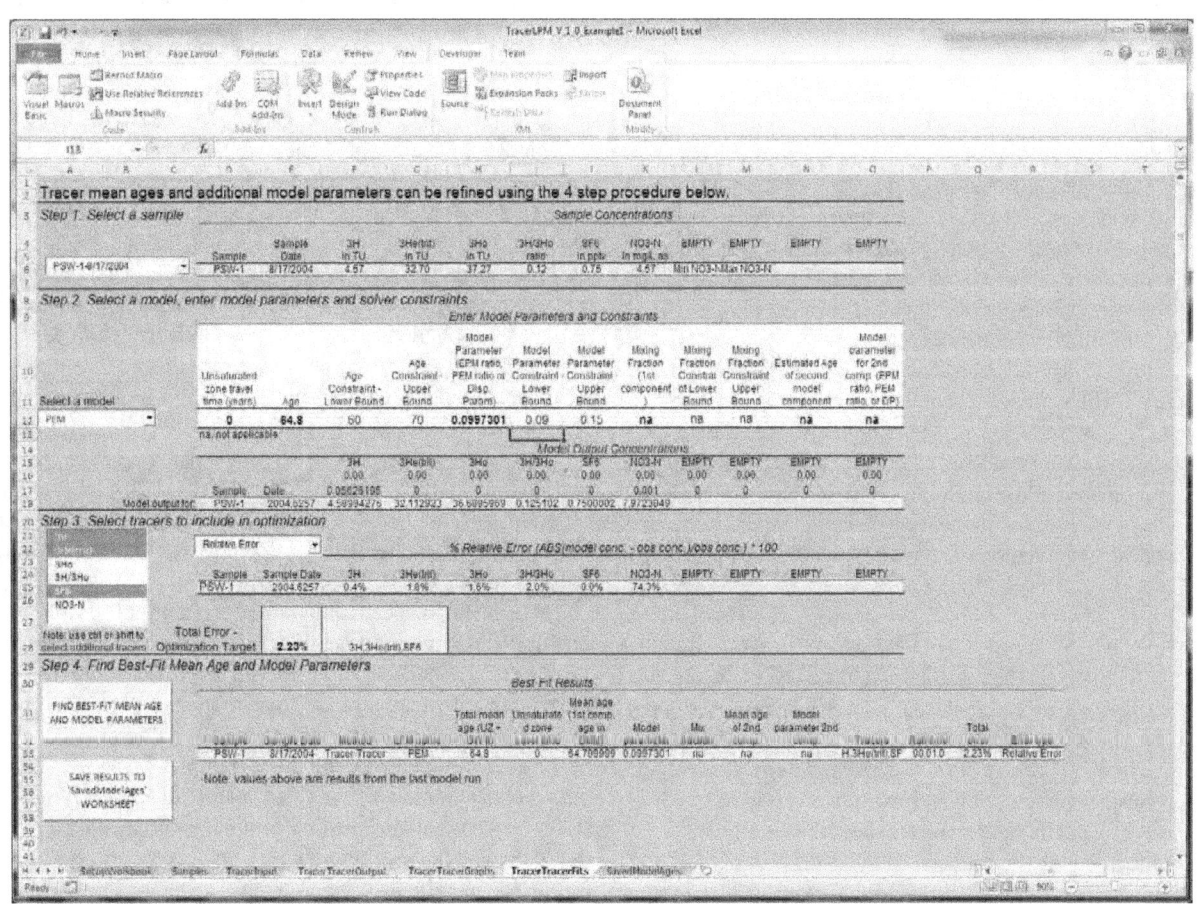

Figure 17. Screenshot of the *TracerTracerFits* worksheet.

Once a model has been defined, the user will need to change the model parameters in bold in row 12. The user should have previously identified an estimated value for the mean age, model parameter, and mixing fraction of the first component, and the mean age and model parameter for the second model component. These values should be specified in row 12 before using the best-fit algorithm. If the user did not specify a BMM, then the mixing fraction, mean age, and model parameter for the second fraction can be ignored.

Users also will need to specify lower and upper bounds for the mean age, model parameter, and mixing fraction. These values are used to constrain the minimization algorithm to finding the optimization between the upper and lower bounds. The lower bounds should be less than or equal to the mean age, model parameter, and mixing fraction for each parameter. The upper bounds should be greater than or equal to the mean age, model parameter, and mixing fraction for each parameter. As a general rule, the upper and lower constraints should be as narrow as possible to minimize the time necessary to complete the calculations. For mean age, the upper and lower constraints should be less than 20 years apart. For model parameters (DP, EPM ratio, PEM ratio) and mixing fractions, the upper and lower constraints should be less than 0.2 units apart. Each of the parameters has at least one absolute boundary that should be followed. The mean age has an absolute lower bound of 0.1 year; therefore, cell E12 should not be lower than 0.1, unless the model is PFM. There is no ceiling to mean age, although most groundwater analyzed for the tracers listed in this workbook generally will be less than 50,000 years old. The model parameters typically have absolute lower bounds of 0.001. A practical upper bound for the EPM ratio and DP is 3. The mixing fraction always will be between 0 and 1.

By default, the difference between measured concentrations and modeled concentrations is measured by relative error. The relative error for each tracer is displayed in row 25. The user also has the option to use the relative squared error as a measure of difference. Users can specify the tracers to be used in the optimization by selecting at least one tracer from the list box on the worksheet. Selection of tracers will cause the summation of relative errors for each tracer in cell E28.

The relative error in cell E28 is minimized by clicking the button named, "FIND BEST-FIT MEAN AGE AND MODEL PARAMETERS" on the lower left-hand side on the worksheet. The program will calculate the optimization error at discrete points for the entire range of mean ages and model parameters specified by their constraints to find the approximate location of the global minimum. Solver, subsequently, is used to refine the mean age and model parameter about this minimum to find the true global minimum. Because the program calculates model output concentrations for the entire range of model constraints, the entire calculation can take several seconds to a few minutes to complete. Once the program

is finished, the results of the best-fit routine are returned to row 33. These results can be stored to the *SavedModelAges* worksheet by clicking the button to the left of the output. This allows the user to model other samples or try different tracer combinations in the best-fit routine.

Time-Series Workgroup

The Time-Series workgroup is used for determining the LPM and mean age by using a single tracer measurement (or multiple tracer measurements) from several samples collected from the same well or spring. The Time-Series workgroup consists of three worksheets: *TimeSeriesOutput*, *TimeSeriesGraphs*, and *TimeSeriesFits*. The tabs of these worksheets are colored blue.

TimeSeriesOutput Worksheet

The *TimeSeriesOutput* worksheet is used to select LPMs and view the model output for multiple sample dates (fig. 18). LPMs are selected from the four pull-down menus in column A. The mean ages and model parameters are defined for each LPM in columns B through H. The selection of LPMs from the first four pull-down menus and changes to model parameters in this worksheet are linked to the models and model parameters in columns G through M on the *TimeSeriesGraphs* worksheet.

Upon selection of a model, tracer output concentrations will be calculated for each sample date listed in column J and each tracer listed in row 2. Up to four models can be selected, and each model has a corresponding set of calculations in columns L through AY. Up to 1,000 sample dates can be modeled from column J. The sample dates listed in column J can be altered manually to suit a more specific range of sample dates. Alternatively, the sample dates in column J can be automatically adjusted by entering a date range in cells A20 and B20, and an increment of time in cells C20. The workbook program will calculate the range of sample dates beginning at the end date and incrementally stepping backward in time until the start date is reached.

If a BMM is specified in one of the four pull-down menus, the mixing fraction, mean age, and model parameter for the second model component is specified in columns F through H. The output tracer concentrations for the LPM of the second component will be calculated in the corresponding set of tracer output in columns AZ through CM. The calculated output concentrations are used as input in the BMM. The tracer concentration of the first component is calculated internally and the resulting tracer concentration of the mixture is returned to columns L through AY.

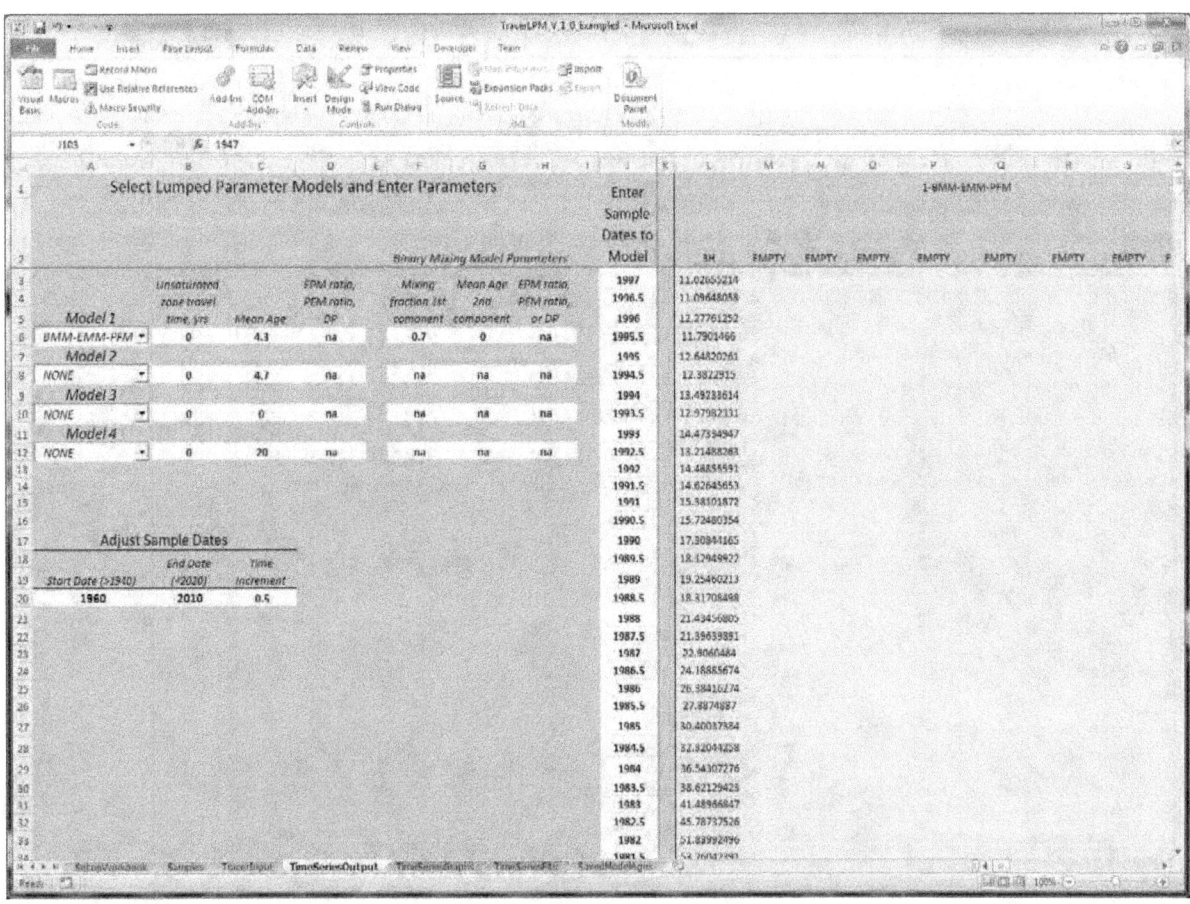

Figure 18. Screenshot of the *TimeSeriesOutput* worksheet.

TimeSeriesGraphs Worksheet

The *TimeSeriesGraphs* worksheet is used to graphically estimate model parameter values for a well or spring where at least one tracer was collected multiple times (fig. 19). The mean age and age distribution for a sample are estimated by viewing tracer output concentrations for various LPMs, mean ages, and other model parameters in relation to the actual measured tracer concentrations from several samples. To populate the graphs, the user will need to select samples, models, and at least one tracer from the pull-down menus beside each graph. Samples can be selected from the list box in the upper left-hand corner of the worksheet.

Lumped parameter models are selected from the pull-down menus in column E. Mean ages and model parameters are defined in the cells to the right of each

pull-down menu. If a BMM is selected from one of the pull-down menus, the mean age and model parameter for the first model component are defined in columns H and I, respectively, and the mixing fraction, mean age, and model parameter of the second model component are defined in columns K through M.

A set of ten graphs are available for the user to create custom time-series graphs. Each graph will display the LPM tracer output concentrations calculated for each sample date in column J of the *TimeSeriesOutput* worksheet. Only samples that have measured tracer values will be plotted on the graphs.

Tracers can be selected from two pull-down menus located to the right of each graph. The pull-down menus contain the tracers selected by the user from the *Samples* worksheet. Each graph has a legend below the x-axis. The models are listed first, and the samples are listed second.

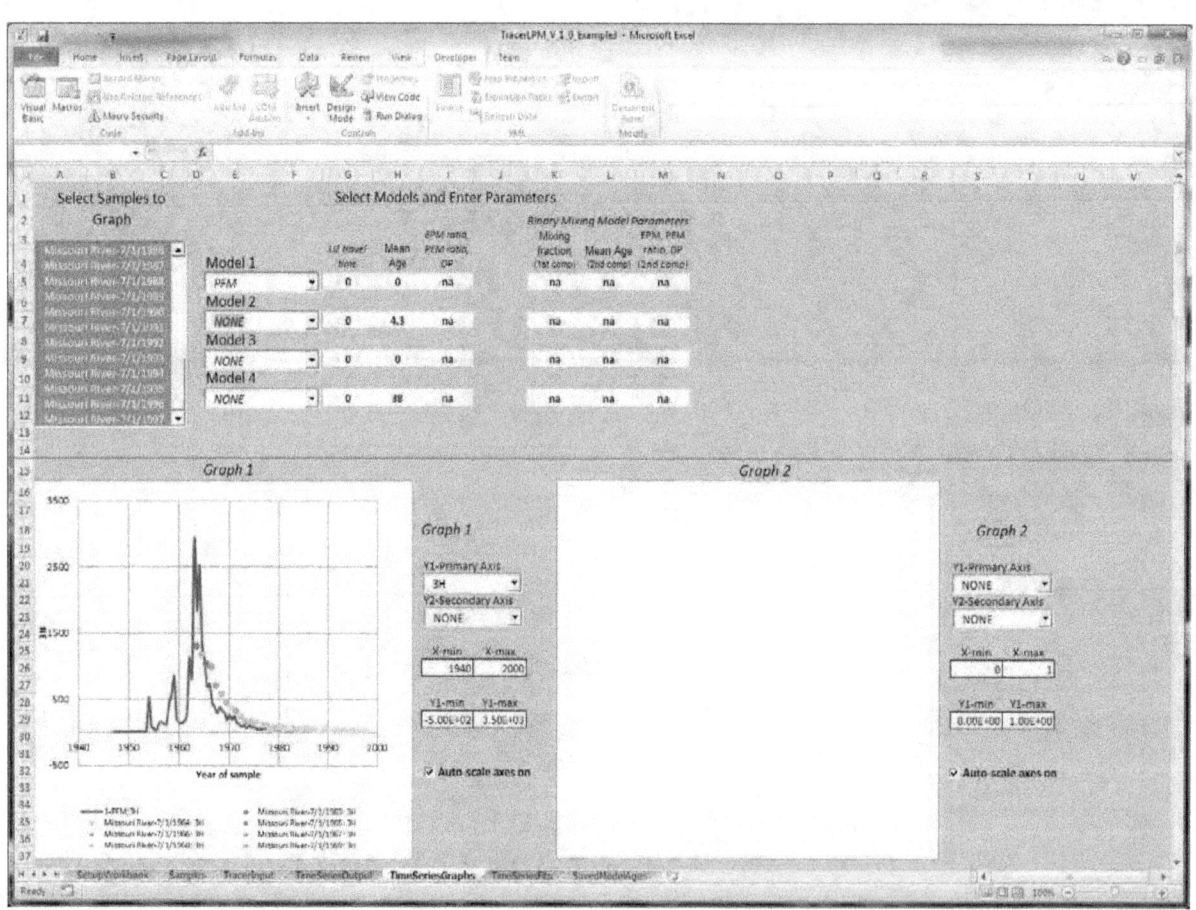

Figure 19.   Screenshot of the *TimeSeriesGraphs* worksheet.

TimeSeriesFits Worksheet

The *TimeSeriesFits* worksheet is used to more accurately determine the mean age, model parameter, or mixing fractions for a well or spring by using time series data (fig. 20 and 21). The best-fit mean age, model parameter, or mixing fractions is found by minimizing the total error between LPM tracer output concentrations and measured concentrations by using a custom search algorithm and Solver (Fylstra and others, 1998).

The user will need to specify at least two samples, a model, and the tracers to use in the optimization. Samples are selected from the list box in the upper left-hand part of the worksheet (fig. 20). There is no limit to the number of samples that can be selected, although the default configuration of the worksheet shows only 10 rows. Selection of more than 10 samples will cause the worksheet to grow vertically, so the row numbers that follow only pertain to the default state. The sample information and measured tracer data will be populated beneath the tracer columns in rows 6 through 15. The model is selected from the pull-down menu in row 20 (fig. 20). The selection of a model causes the program to

calculate tracer output concentrations for each tracer in rows 26 through 35 by using the model parameters in bold from row 20. If the model is a BMM, the user will need to specify the mean age and model parameter for the second model component. For BMMs, the mean age and model parameter for the second model component are entered in cells N20 and O20, respectively. The output concentrations of each tracer in the second component of the BMM are in rows 36 through 45. These concentrations are the input concentrations for the first component in the BMM. These rows are normally hidden from view to prevent confusion with the final output concentrations in rows 26 through 35.

It should be noted that for BMMs, only the first model component is optimized, and the second model component is left static. It is expected that one of the model components is well enough constrained so that the other model component and mixing fraction can be optimized. It is also possible optimize the entire BMM by optimizing one component, then reversing the BMM so that the second component is the first and optimizing that model. This method can require multiple iterations.

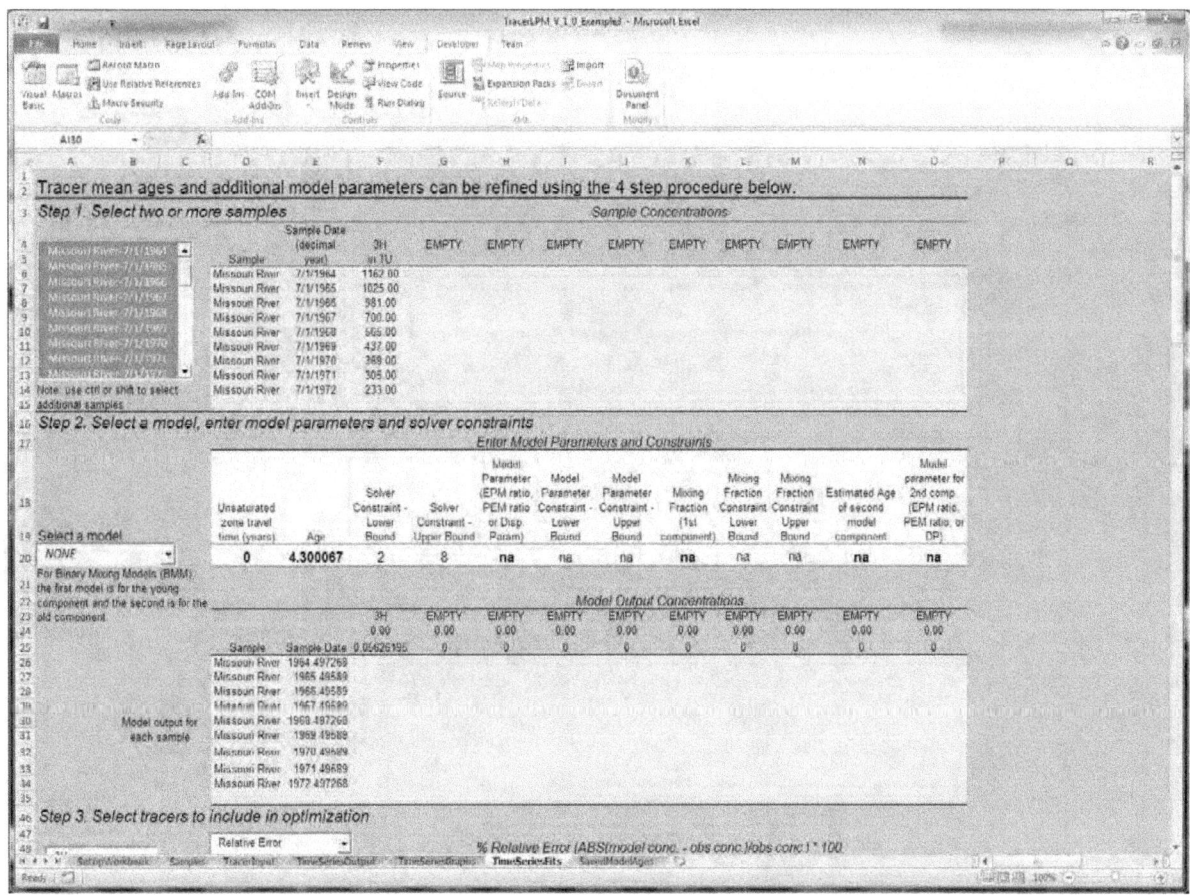

Figure 20. Screenshot of the upper-half of the *TimeSeriesFits* worksheet.

Once a model has been defined, the user will need to change the model parameters in bold in row 20 (fig. 20). The user should have previously identified an estimated value of mean age, model parameter, and mixing fraction for the first component, and mean age and model parameter for the second component. These values should be specified in row 20 before using the best-fit algorithm. If the user did not specify a BMM, then the mixing fraction, mean age, and model parameter for the second model component can be ignored. The user will also need to specify lower and upper bounds for the mean age, model parameter, and mixing fraction. These values are used to constrain the algorithm to finding the optimization between the upper and lower bounds. The lower bounds should be less than or equal to the mean age, model parameter, and mixing fraction for each parameter. The upper bounds should be greater than or equal to the mean age, model parameter, and mixing fraction for each parameter.

As a general rule, the upper and lower constraints should be as narrow as possible to minimize the influence of local minimums on the solution. For mean age, the upper and lower constraints should be less than 20 years apart. For model parameters (DP, EPM ratio, PEM ratio) and mixing fractions, the upper and lower constraints should be less than 0.2 units apart. Each of the parameters has at least one absolute boundary that should be followed. The mean age has an absolute lower bound of 0.1 year; therefore, cell E20 should not be lower than 0.1. There is no ceiling to mean age, although most groundwater analyzed for the tracers listed in this workbook generally will be less than 50,000 years old. The model parameters generally have lower bounds of 0.001. A practical upper bound for the EPM ratio and DP is 3. The mixing fraction always will lie between 0 and 1.

By default, the difference between measured concentrations and modeled concentrations is measured by relative error. The relative error for each tracer is displayed in rows 51 through 60 (fig. 21). The user also has the option to use the relative squared error as a measure of difference. Users can specify the tracers to be used in the optimization by selecting at least one tracer from the list box on the worksheet. Selection of tracers will cause the summation of relative errors for each tracer in cell D63.

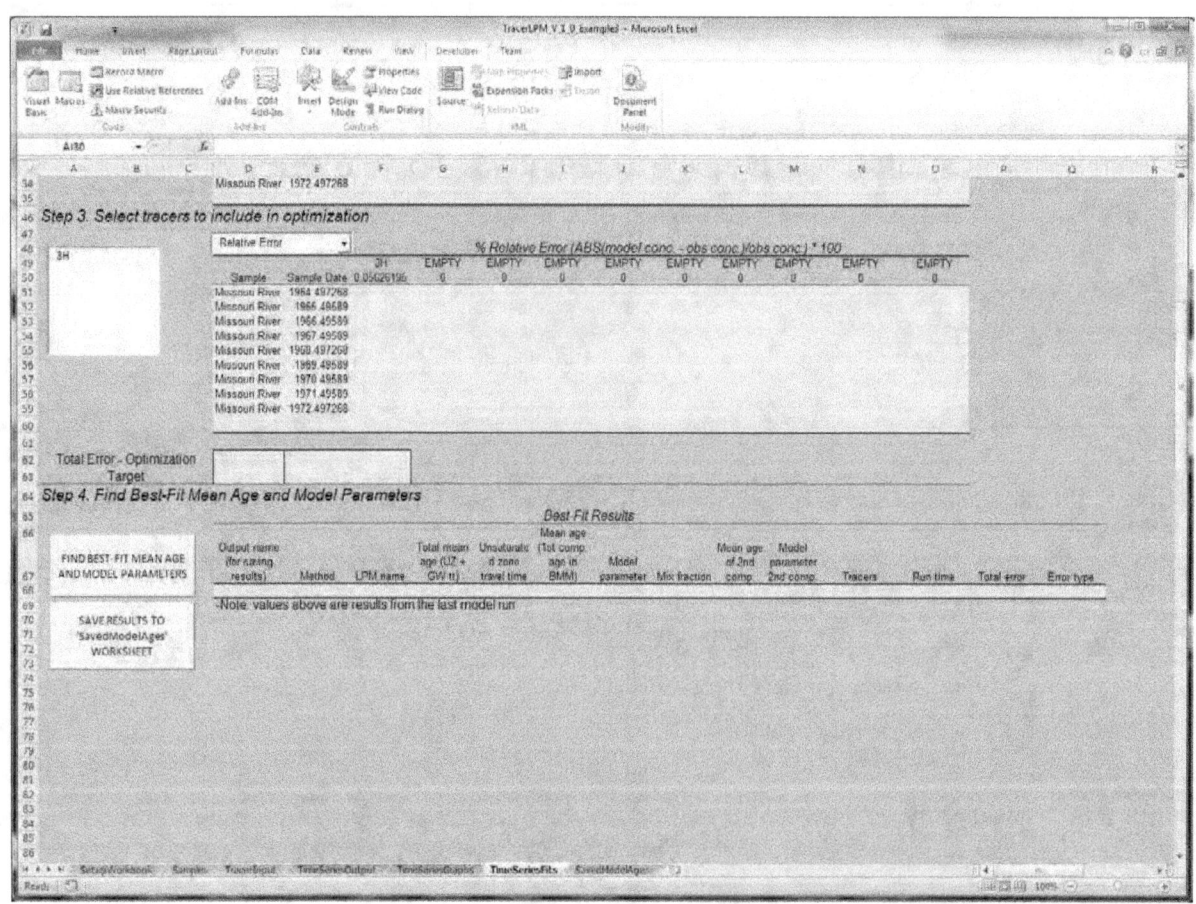

Figure 21. Screenshot of the lower-half of the *TimeSeriesFits* worksheet.

The relative error in cell D63 is minimized by clicking the button, "FIND BEST-FIT MEAN AGE AND MODEL PARAMETERS," on the lower left-hand side on the worksheet (fig. 21). The program will calculate the optimization error for the entire range of mean ages and model parameters specified by their constraints to find the approximate location of the global minimum within the constraints. Solver, subsequently, is used to refine the mean age and model parameter about this minimum to find the true global minimum within the constraints. Because the program calculates output for the entire range of model constraints, the entire calculation can take several seconds to a few minutes to complete. Once the program is finished, the results of the best-fit routine are returned to row 68. These results can be stored to the *SavedModelAges* worksheet by clicking the button to the left of the output (fig. 21). This allows the user to model other samples or try different tracer combinations in the best-fit routine.

SavedModelAges Worksheet

The *SavedModelAges* worksheet is used to store the best-fit results obtained from the *TracerTracerFits* and *TimeSeriesFits* worksheets (fig. 22). These two worksheets will populate the mean age, model parameters, mixing fractions, and other information determined by the best-fit algorithm contained within those worksheets to the *SavedModelAges* worksheet. The best-fit mean ages and model parameters can be used to keep track of models used to report the mean age of samples and for reference when graphing the age distributions of those models or forecasting future concentrations. On this worksheet, rows can be deleted or cleared to remove previous results after row 3.

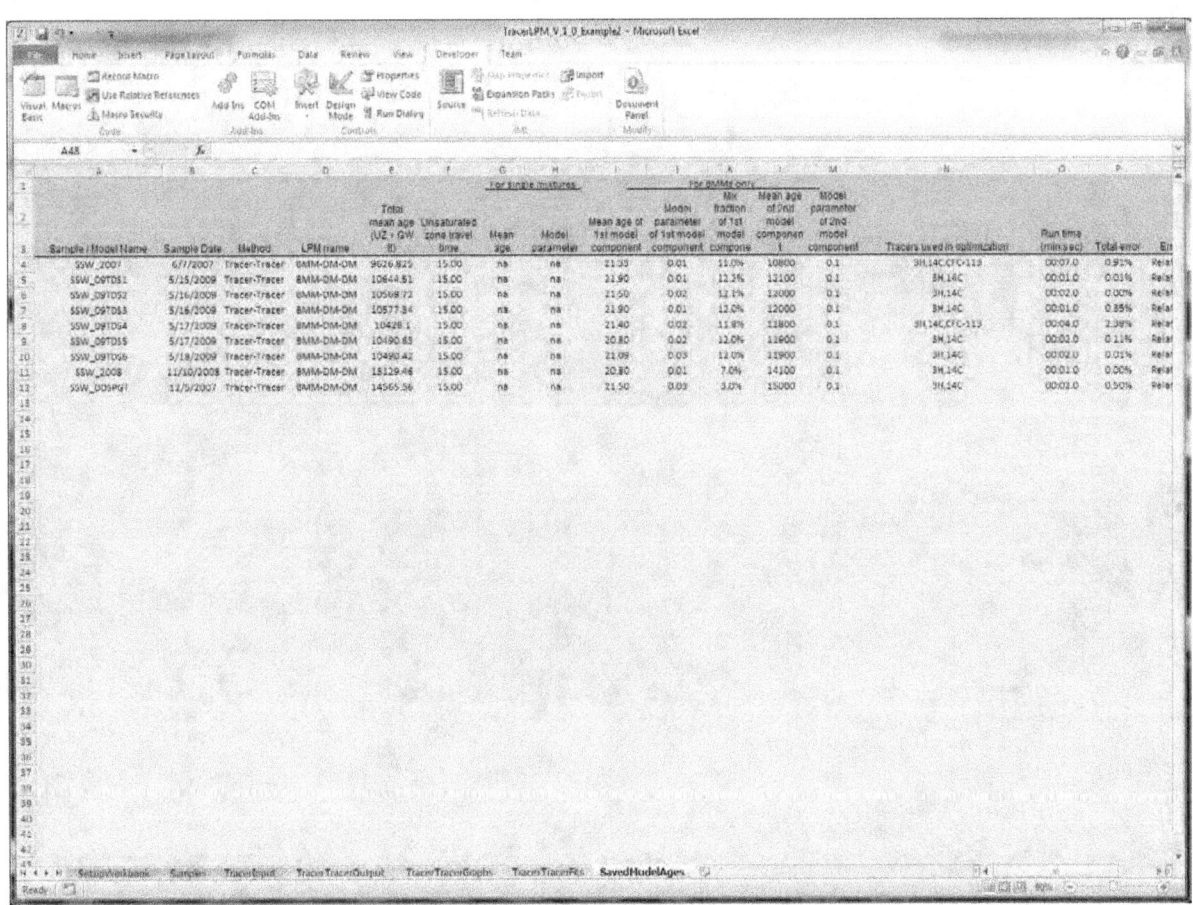

Figure 22. Screenshot of the *SavedModelAges* worksheet. This worksheet is populated from the *TracerTracerFits* and *TimeSeriesFits* worksheets.

Age Distribution & Forecasting Workgroup

The Age Distribution & Forecasting workgroup is used for viewing user defined age distributions or the age distributions of LPMs, and for predicting past and future concentration trends of non-point-source contaminants at a well or spring for multiple management scenarios. The Forecasting workgroup consists of three worksheets: *UserDefinedAge*, *LPM_AgeDistribution* and *Forecasting*. The tabs of these worksheets are colored purple.

UserDefinedAge Worksheet

The *UserDefinedAge* worksheet can be used to enter up to four user-defined age distributions (fig. 23). The age distributions are entered in columns B through P. For each age distribution, the ages and corresponding fractions are entered into the white, non-filled cells. Clicking the button, "GRAPH AGE DISTRIBUTIONS (REFRESH GRAPHS)," will cause the program to calculate the cumulative fraction in the blue columns next to each age distribution, and graph the frequency and cumulative distributions (fig. 23). The program will also calculate the mean age, total recharge fraction, and the minimum age interval of each age distribution. The total recharge fraction should be 1, or very nearly 1, in order to calculate tracer concentrations and mean age from the age distribution accurately.

Tracers selected from the *Samples* worksheet will be included in the list of tracers in column S. Clicking the button, "CALCULATE TRACER OUTPUT," will cause the program to calculate the tracer output concentrations in columns V through Y for the sample date and unsaturated travel time specified in rows 5 and 6 for each age distribution.

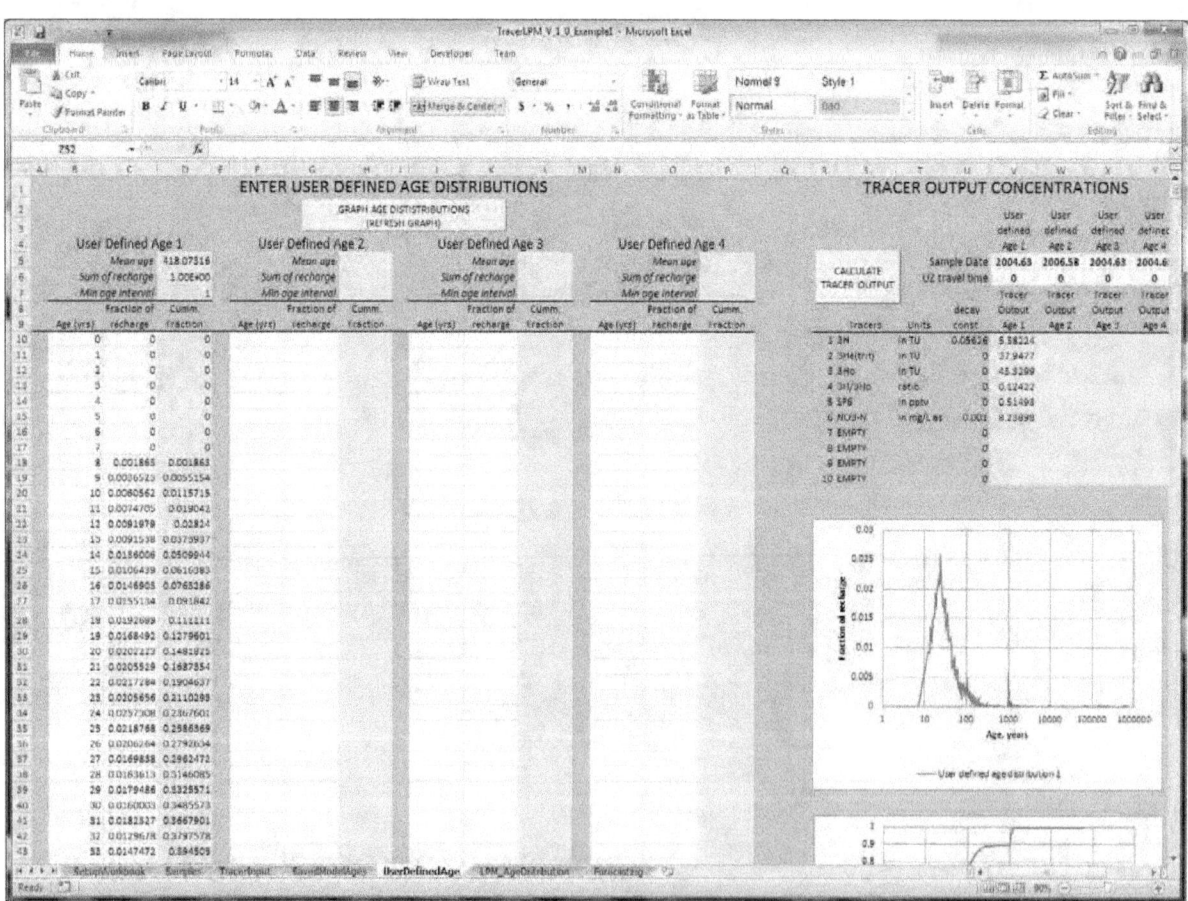

Figure 23. Screenshot of the *UserDefinedAge* worksheet.

LPM_AgeDistribution Worksheet

The *LPM_AgeDistribution* worksheet is used to view the exit age distribution [*g(t)*] of the LPM(s) determined for samples by using the tracer-tracer or time-series method (fig. 24). The age distribution shows the fractional contribution of each sub-parcel of water that collectively compose the whole sample. The program automatically calculates the cumulative fraction for each age distribution in the adjacent column and is graphed in the chart to the right of the age-frequency curve (not shown in fig. 24). The fraction of sample is the total fraction of the age interval integrated from the previous age (current age minus time step) to the current age listed in column A (rows greater than 17). The average age of each interval is listed in column B. Up to four models can be viewed simultaneously by selecting a model from each pull-down menu in columns D, G, J, and M. User defined age distributions that were entered on the UserDefinedAge worksheet can also be selected and viewed alongside any LPM age distribution.

After selection of an LPM, the mean age and any additional model parameters can be changed in rows 4 and 5. For BMMs, the mixing fraction, mean age of old water, and model parameter for the old model are specified in rows 9 through 11, respectively. Each age-distribution function specified in row 3 will be calculated at each time step for the total number of years specified in cells A14 and A16. These two values can be changed to suit user needs; however, the number of calculations is proportional to the total number of years, divided by the time step. As such, increases in the total number of years and shorter time steps increase the time between calculations.

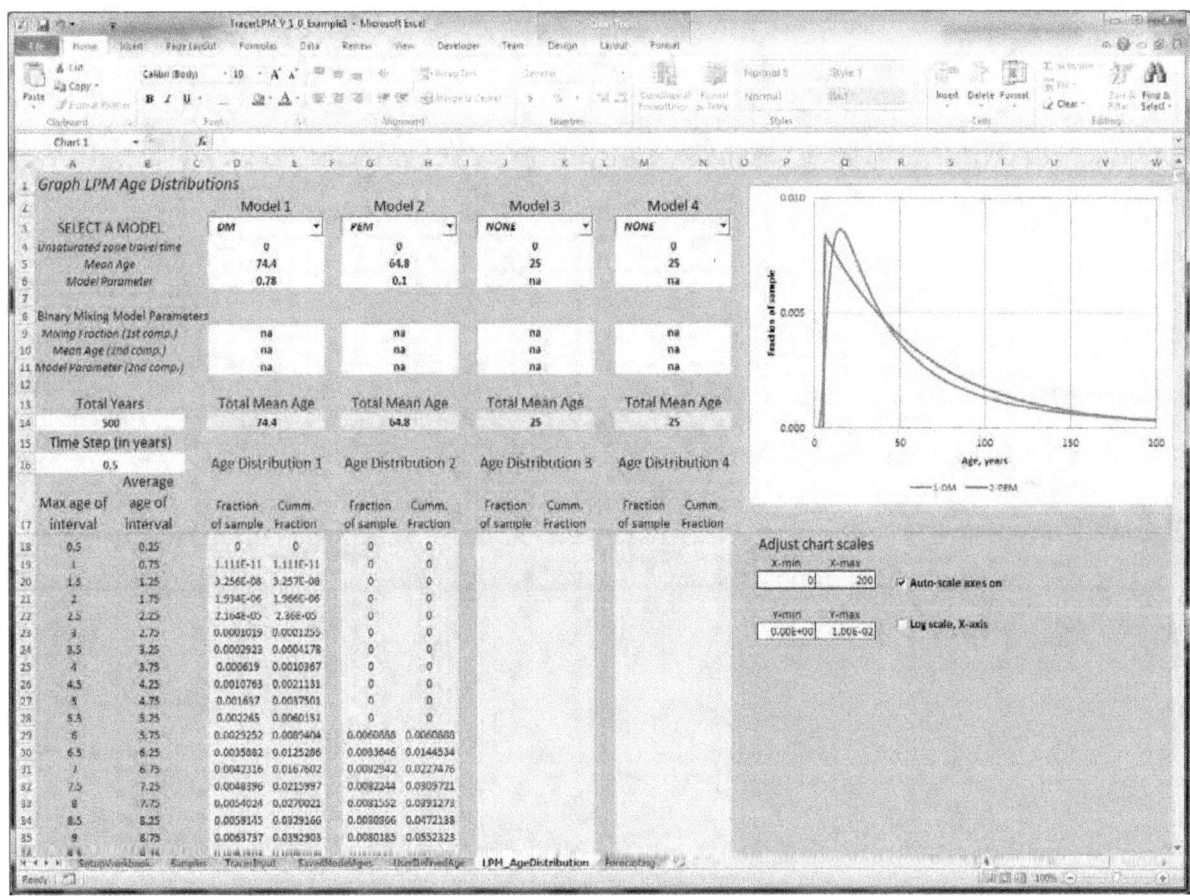

Figure 24.   Screenshot of the *LPM_AgeDistribution* worksheet. The cumulative age distribution is plotted on a graph to the right of the age frequency plot and is not shown here.

Forecasting Worksheet

The *Forecasting* worksheet is used to view output of the LPMs or user-defined age distributions to forecast future concentrations for different tracer input scenarios, assuming the tracer concentrations are uniformly distributed across the entire recharge area (fig. 25). The dates listed in column A correspond to the concentration of the tracer in recharge (the first column beneath each scenario) and to the model tracer concentration expected from the well or spring, if sampled during that year (the second column beneath each scenario).

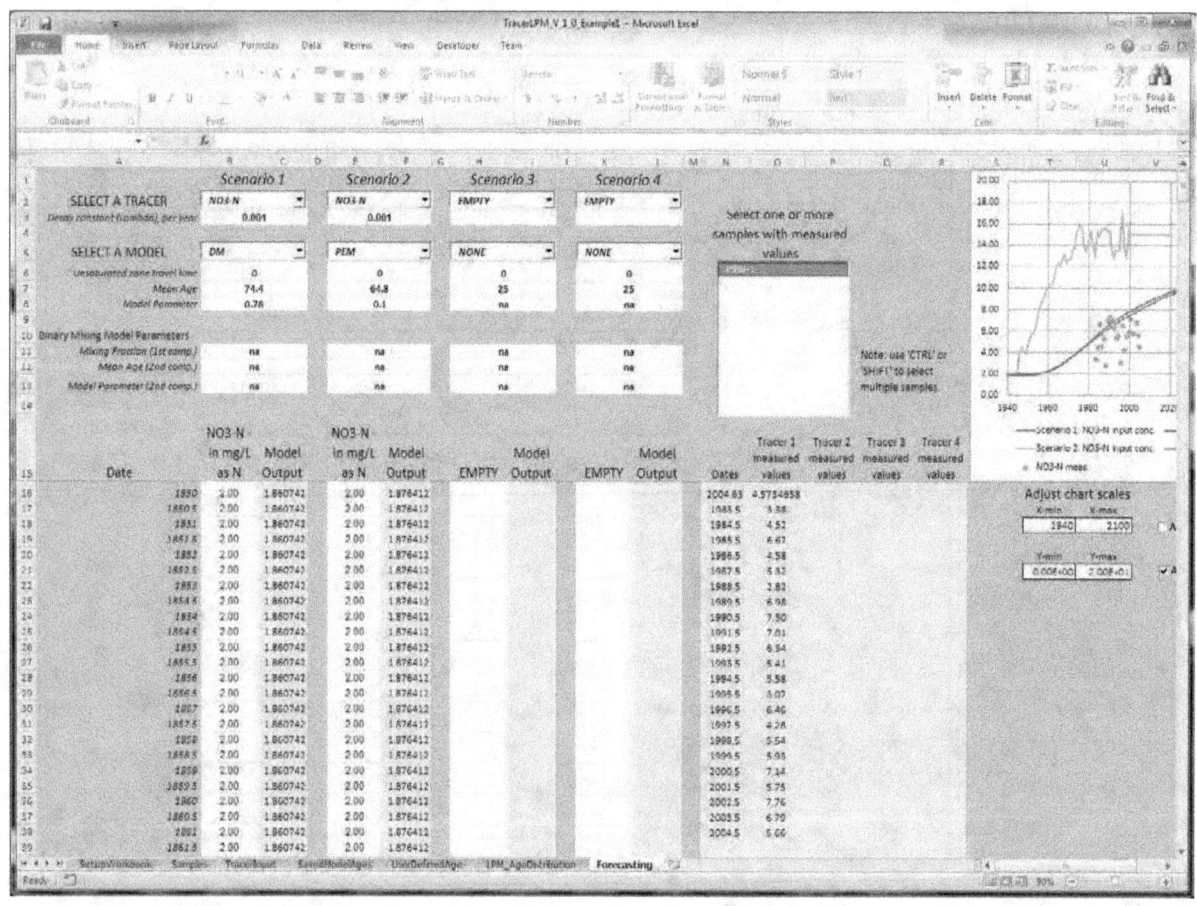

Figure 25. Screenshot of the *Forecasting* worksheet.

To begin using the worksheet, a tracer must be selected from the pull-down menus in row 2. Selecting a tracer will cause the program to retrieve stored tracer input history and add it to the first column below the pull-down menu (columns B, E, H, or K) beginning in row 16. Because tracer input data are not stored beyond year 2020, the user will need to define the input history for years after 2020 in order to forecast concentrations in the well or spring beyond 2020. The input history can be manipulated to simulate constant input, cessation, decreasing, or increasing inputs of the tracer concentrations in recharge. As a result, the response of the well or spring to changes in the input history can be evaluated for different scenarios.

Select a model or user-defined age distribution from one of the four pull-down menus to calculate and graphically view tracer concentrations in a well or spring on the basis of the tracer input scenario devised shown in figure 25. The mean age and any additional parameters for the selected model are specified below the pull-down menus. The mean age, additional model parameters, and mixing fractions for the BMMs can be defined in rows 11 through 13, respectively.

Water-quality or tracer data that were entered on the *Samples* worksheet can be added to the graph on the *Forecasting* worksheet by selecting one or more samples from the list box to the left of the graph (fig. 25). The data entered on the *Samples* worksheet should have the same units of measure as the tracer input data. Selecting a sample name from the list box will populate all numeric sample dates and tracer data for that sample name in columns N through R. The tracer data will automatically be displayed in the graph if the check box below the graph is checked. The historical water-quality data can be removed by clearing the check box below the chart.

Examples

TracerLPM can be used for various purposes such as viewing tracer-tracer and time-series plots, evaluating mixing models and age distribution of samples, detecting anomalous data and local processes affecting tracer concentrations, and calculating past and future concentrations of non-point-source contaminants in wells or other groundwater receptors. In this section, Example 1 shows how anthropogenic atmospheric tracers were used to determine the mean age and age distribution of a sample from a public-supply well in Modesto, California, and how an LPM calibrated to environmental tracer data can be used to evaluate historical and future responses of the well to changing nitrate concentrations in recharge (TracerLPM_V_1_Example1.xlsm). Example 2 shows how ^{14}C was used with other tracers to determine the mean age and age distribution of mixed water, having a large range of ages and multiple sources of recharge, from a public-supply well in Albuquerque, New Mexico (TracerLPM_V_1_Example2.xlsm). Example 3 shows how a long-term record of tritium data from the upper Missouri River was used to determine the residence time of water in the river basin and how the calibrated LPM can be used to evaluate past and future responses of a river to changing nitrate input in the watershed (TracerLPM_V_1_Example3.xlsm).

Example 1: Public-Supply Well in Modesto, California

A public-supply well in the Central Valley principal aquifer in Modesto, California, was studied as part of the NAWQA program's "Transport of Anthropogenic and Natural Contaminants (TANC) to supply wells" topical team (Eberts and others, 2005; Burow and others, 2008; Jurgens and others, 2008). The well is located in a suburban neighborhood that is close to urban as well as agricultural sources of contamination. The aquifer is unconfined in the upper part and becomes semi-confined at depths greater than 45 meters (m), or about 150 feet (ft), below land surface. The aquifer is a heterogeneous mixture of clay, sand, silt, and gravel. Recharge to the aquifer predominantly comes from excess irrigation infiltration beneath agricultural land that surrounds the city, as well as minor amounts of lawn-watering within the urban

landscape, and precipitation. Depth to water at the well is about 9 m (30 ft), and the well is screened from about 27.4 to 111 m (90 to 365 ft).

A sample of water from the well was collected in August 2004 and analyzed for concentrations of ^{3}H, ^{3}He$_{trit}$, and SF$_6$. The mean age of this sample was determined by modeling ^{3}H, ^{3}He$_{trit}$, ^{3}H$_o$ (^{3}H + ^{3}He$_{trit}$), ^{3}H/^{3}H$_o$, and SF$_6$ by using the TracerLPM workbook. Values of ^{3}H, ^{3}He$_{trit}$, ^{3}H$_o$, ^{3}H/^{3}H$_o$, and SF$_6$ were 4.57 TU, 32.70 TU, 37.27 TU, 0.123 (ratio), and 0.75 pptv, respectively.

The LPM and mean age of the sample were evaluated by using the Tracer-Tracer method because multiple tracers were collected on a single sample date. This was done by first comparing the sample tracer concentrations to the output tracer concentrations of the PFM and EMM by using the setup shown in figure 26. For all tracer combinations, the PFM does not compare well with the measured tracer

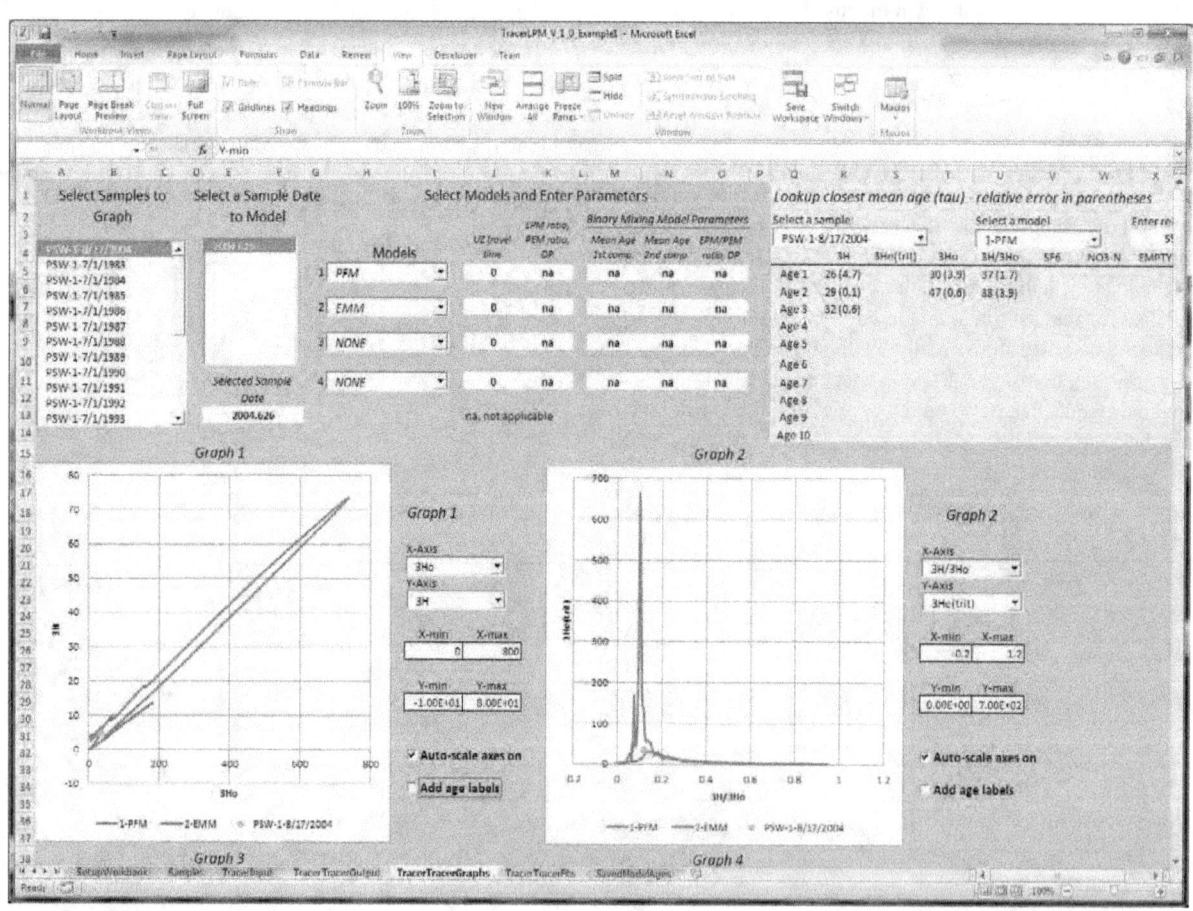

Figure 26.   Screenshot of set-up of *TracerTracerGraphs* worksheet showing the selection of the PFM and EMM for a sample collected on August 17, 2004.

concentrations, which indicates that the sample represents a mixture of groundwater ages, and the PFM model should not be considered in the analysis of mean age for this well (fig. 27). The EMM is consistent with the $^3H/^3H_o$ and SF_6 values of the sample, but the observed concentrations for other tracer values deviate from the modeled concentrations, indicating other models should be considered. The PEM and DM were evaluated because these tracer concentrations lay between the EMM and PFM curves. In addition, the proximity of ^3H, $^3He_{trit}$, and ^3H_o to the EMM graph indicates that the optimal parameter values for the PEM and DM are likely to be representative of a more mixed system than one dominated by piston-flow. Although the EPM can be used to model tracer concentrations, the PEM was chosen instead because the well and aquifer most resemble the configuration depicted by the PEM (fig 4).

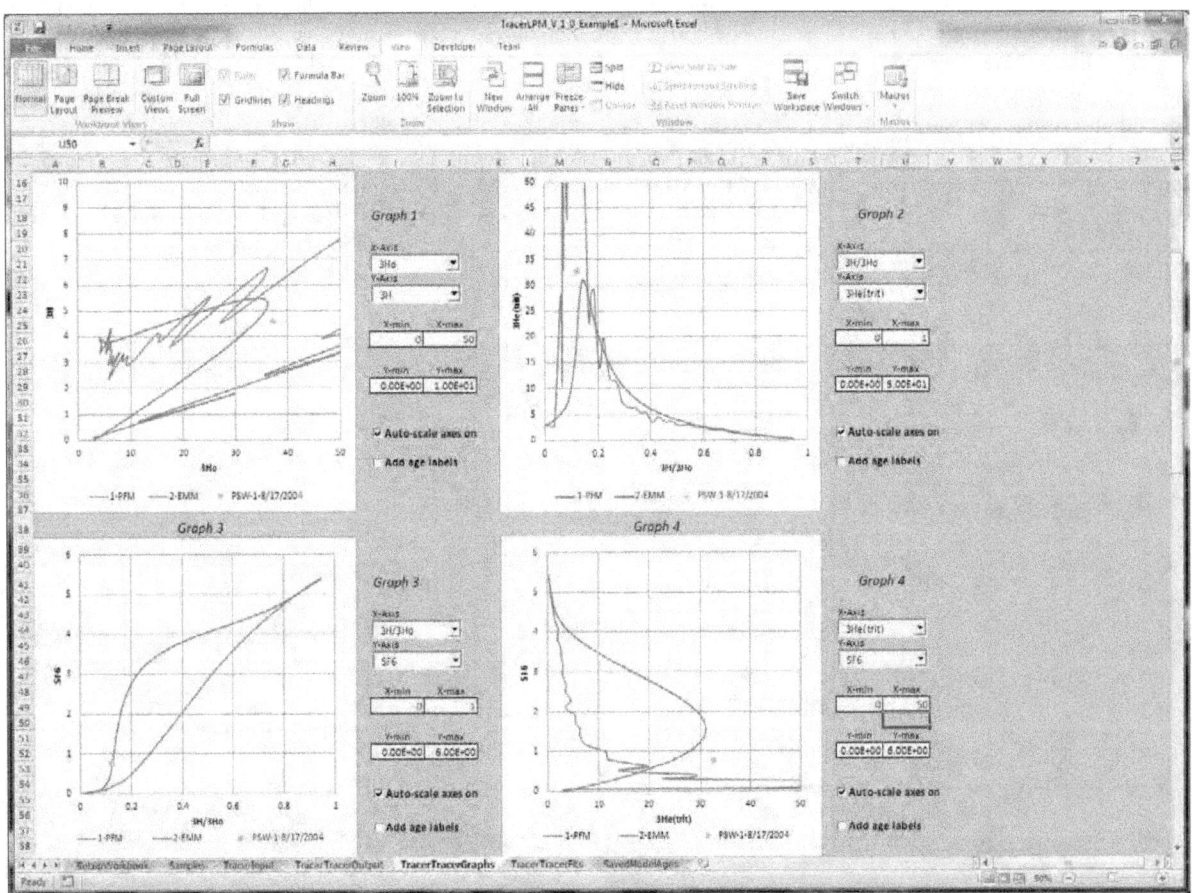

Figure 27. Tracer-Tracer graphs of PFM and EMM output concentrations in comparison to measured concentrations for a sample collected on August 17, 2004.

The PEM and DM were in good agreement with all five tracer values by using parameter values of 0.1 and 0.8, respectively (fig. 28). Estimated mean ages using these values were between 60 and 80 years (see Lookup Mean Ages section). These estimates of mean age and parameter values then were used to fit the measured concentrations with the modeled concentrations by using the *TracerTracerFits* worksheet. The PEM yielded an optimized mean age of 64.8 years with a PEM ratio of 0.10, and the total error between modeled values and observed tracer values was about 2.3 percent (for ^{3}H, ^{3}He$_{trit}$, and SF$_6$ only). The DM yielded an optimized mean age of 75.3 years with a dispersion parameter (*D/vx*) of 0.8. The total error between modeled values and observed tracer values was 4.4 percent (for ^{3}H, ^{3}He$_{trit}$, and SF$_6$ only). Both models had total errors less than 10 percent, which indicates both models fit the measured tracer concentrations well.

In a homogeneous undisturbed system, the PEM ratio should be close to the ratio of the unsampled portion to the sampled portion of the aquifer. In some cases, it could be possible to use well construction information to inform an initial estimate of the PEM ratio for this model. In the Modesto example, a hypothetical PEM ratio of 0.22 was calculated by using the water level and well-construction information provided previously. The hypothetical ratio is similar to the one determined from fitting tracer data (0.10), but the difference possibly indicates that a younger, more shallow fraction of water is present in the sample than would be predicted from the hypothetical PEM ratio. This could indicate that heterogeneity has allowed faster transport of tracers to the well than would be expected from homogeneous conditions or that pumping has drawn younger water downward to the well (i.e., the vertical age distribution of water entering the well does not conform to the exponential profile of the PEM model). Most of the error in the fit is from SF$_6$, which is not modeled well. If SF$_6$ is removed from the fitting routine and the model is rerun, the mean age of the sample is 67 years with a PEM ratio of 0.13.

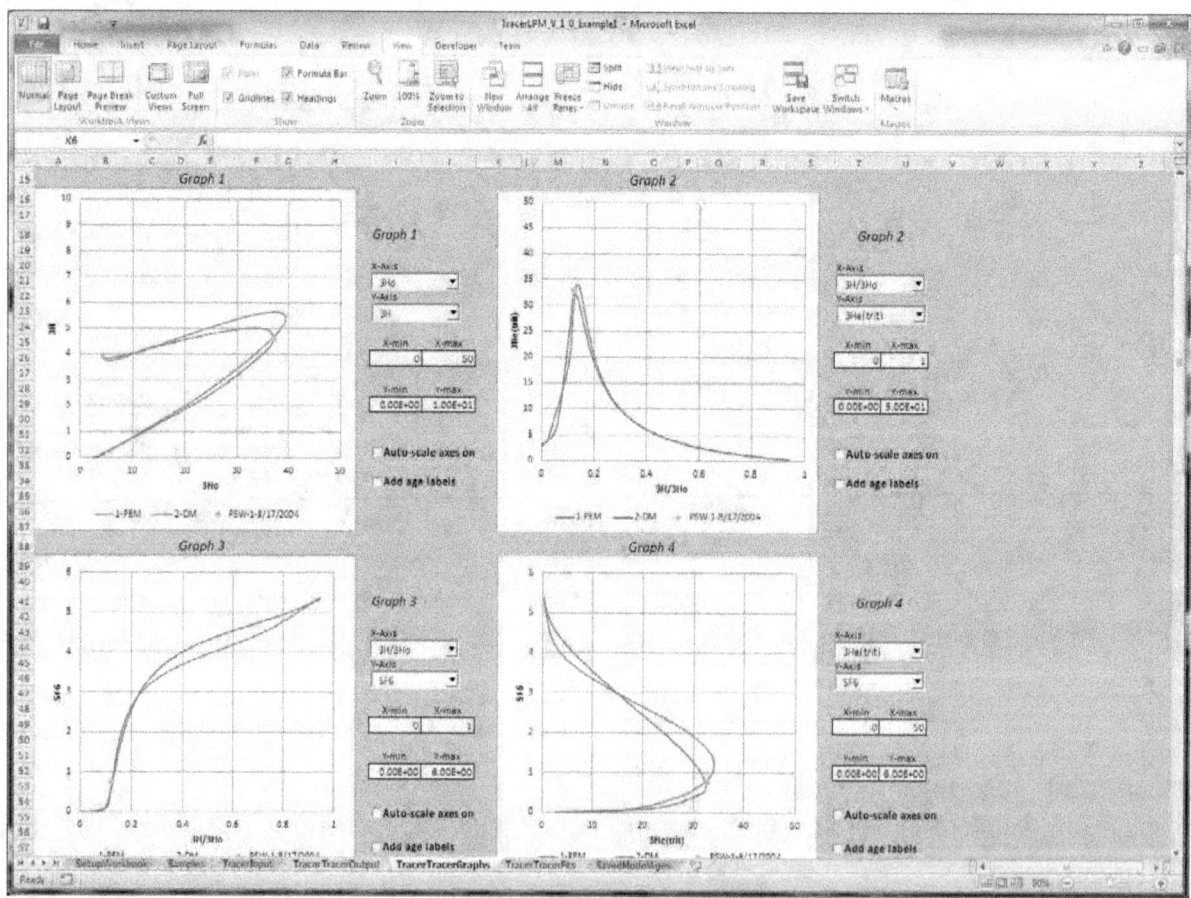

Figure 28. Tracer concentrations for the Modesto, California, sample collected on August 17, 2004 (solid green circle), and tracer concentrations for various mean ages modeled by the partial exponential model (solid blue line) and dispersion model (solid red line). The partial exponential model has a PEM ratio of 0.1, and the dispersion model has a dispersion parameter of 0.8.

Overall, the PEM and DM have similar age distributions, and both indicate that water contributing to the well has a broad distribution of ages (fig. 29). All three models (PEM, DM, and particle tracking) indicate relatively little recently recharged (few years or less) groundwater in the well, reflecting the fact that the top of the screen is some distance below the water table. The models indicate that the amount of young water (less than 50 years) reaching the well is greater than 50 percent (56 percent for the DM, and 53 percent for the PEM) and about 20 percent of the water from this well is older than 100 years.

The age distributions determined from the lumped parameter models were similar to the age distribution determined from a groundwater flow model with advective particle tracking (Burow and others, 2008; Eberts et al., 2012). The groundwater flow model was calibrated against water-level, SF_6, and ^3H data from the public-supply well and 18 monitoring wells. These results suggest the LPMs can approximate the age distribution of the well (fig. 29). Both LPMs show a larger mass of younger water than

the particle-tracking age distribution, which has a peak at 24 years. Tracer concentrations calculated from the particle-tracking age distribution were 5.38 TU, 37.9 TU, 43.3 TU, 0.124 (ratio), and 0.50 pptv for ^3H, $^3He_{trit}$, ^3H_o, $^3H/^3H_o$, and SF_6, respectively. The total relative error between measured and modeled concentrations was 67 percent (for ^3H, $^3He_{trit}$, and SF_6, only), which is larger than the total error from the LPMs. The mean ages reported for this example were similar but slightly older than the ages reported by Eberts and others (2012), who found mean ages of 54 and 59 years for the EPM and DM, respectively, and an EPM ratio of 0.2 and a dispersion parameter of 0.51. In that analysis, the global minimum error between measured and modeled tracer concentrations was determined by a combination of manual calibration from the inspection of tracer-tracer graphs and Solver. Slight differences in tracer input data, and differences in the discretization of time for model calculations (TracerLPM uses monthly time-steps) and in the discretization of parameter increments for the search of optimal parameters, were responsible for most of the discrepancy in results.

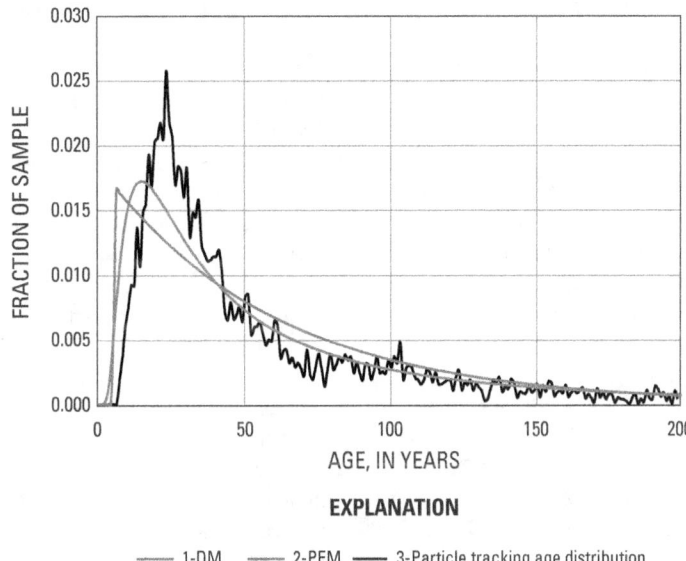

EXPLANATION

—— 1-DM   —— 2-PEM   —— 3-Particle tracking age distribution

Figure 29. Modeled age distribution of a sample collected from a public-supply well in August 2004, Modesto, California.

The models determined from the age tracers can be evaluated for their accuracy and predictive capability by comparing the output of water-quality constituents where the input function is known or estimated to the actual historical water-quality data. McMahon and others (2008b) used the age distribution determined from particle tracking to forecast and compare historical nitrate (NO_3^-) concentrations at the Modesto, California, public-supply well. Nitrate is an excellent constituent to evaluate the models because it has been analyzed at this well 129 times since 1966 (Jurgens and others, 2008), and Burow and others (2008) estimated the input function for nitrate in groundwater beneath agricultural areas near Modesto by using historical nitrogen sales. The input of nitrate from the application of nitrogen fertilizers is the most important source of nitrate in groundwater in the Modesto area. Nitrate was added to the list of tracers in the workbook, and the decay rate for nitrate (by denitrification) in Modesto groundwater was estimated to vary from 0 to 0.02 milligrams per liter (mg/L) as nitrogen (N) per year (McMahon and others, 2008a). A decay rate of 0.001 per year was used in the model presented here.

Figure 30 shows the response of nitrate by using the PEM (solid red line), DM (solid blue line) and particle-tracking age distribution (dashed black line; from Burow and others, 2008), along with historical nitrate concentrations from the well. The input concentration for nitrate was assumed to be constant beginning in 2000 until the year 2020, when nitrate concentrations were set to zero, simulating cessation of nitrate input into the groundwater system. The results are similar to results published by McMahon and others (2008b), although the input history for nitrate after 2005 was different in their simulation. The results show that both the PEM and DM predict very similar behavior of nitrate at the well and, generally, agree with the particle tracking results. All three models predict nitrate concentrations in the well could continue to increase while inputs remain constant as older groundwater components in the sampled mixture approach steady state with the input concentrations. Recall that the models (LPMs and the particle-tracking age distribution) assume tracer concentrations are distributed uniformly across the recharge area. In reality, nitrate concentrations are not likely to be distributed uniformly because the capture zone of this well includes areas of urban landscape that have relatively little nitrate loading and areas of agricultural landscape that have higher nitrate loading.

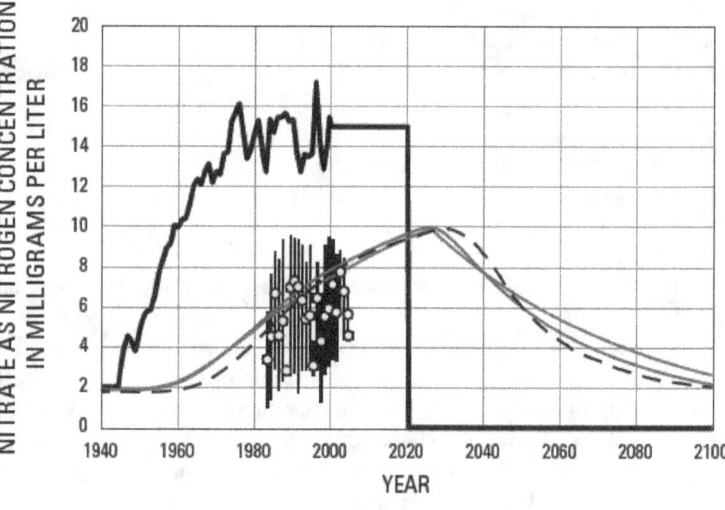

Figure 30.   The simulated response of nitrate (NO_3) by using the PEM (soilid red line), DM (solid blue line), and a user-defined age distribution determined from a groundwater flow model and particle tracking (dashed black line; Burow and others, 2008). The solid black line shows the input concentration of nitrate at the water table (Burow and others, 2008) through 2000, followed by a hypothetical future scenario, including a decrease to 0 after 2020. The yellow circles are measured nitrate concentrations from the public-supply well (Jurgens and others, 2008).

The effect of spatial variations of land-use on nitrate loading in recharge was considered in McMahon and others (2008b). The particle-tracking approach has an advantage compared with LPMs because location information can be associated with the age distribution for building more realistic tracer-input functions. Nevertheless, it is clear that the LPMs can predict similar behavior to age distributions obtained from particle tracking. The LPMs predict similar peak concentrations and decline of nitrate concentration around the year 2026, whereas the particle tracking age distribution predicts a slightly later peak around 2030. In addition, all three models predict a continuation of higher concentrations after application of nitrate has ceased because the youngest fractions are missing from age distributions of the well.

Example 2: Public-Supply Well in Albuquerque, New Mexico

A public-supply well located within the Rio Grande principal aquifer in Albuquerque, New Mexico, was studied as part of the NAWQA program's "Transport of Anthropogenic and Natural Contaminants (TANC) to supply wells" topical team (Bexfield and others, 2012). Samples for ^{3}H, ^{14}C, and chlorofluorocarbons (CFCs) were collected in June and December 2007, November 2008, and May 2009. The data in this report are from the studied supply well or SSW in Bexfield and others (2012). All ^{14}C measurements were corrected for dilution from non-radioactive ("dead") carbon sources by using PHREEQC and NetpathXL (Parkhurst and Appelo, 1999; Plummer and others, 1994). CFC aqueous concentrations were adjusted for equilibrium with the atmosphere and excess air prior to entering their concentrations into TracerLPM.

The main source of natural recharge for groundwater in this area is the Rio Grande. In the vicinity of the well, the aquifer is dominated by old groundwater, which has ^{14}C ages ranging from about 4,500 years in the shallow part of the aquifer, approximately the upper 15 m (50 ft) below the water table, to more than 20,000 years in the deep part of the aquifer, which is generally lower than 76 m (250 ft) below the water table. A relatively thick unsaturated zone, generally ranging from 45 to 76 m (150 to 250 ft) thick, and low recharge rates from the land surface indicate the aquifer should have very little young groundwater; however, ^{3}H, ^{3}He$_{trit}$, and CFCs commonly are detected in the shallow and intermediate depths of the aquifer (Bexfield and others, 2012). Possible sources of relatively recent recharge to the groundwater system could be seepage from the Rio Grande or leakage from buried water-supply distribution lines. The supply well is screened across the intermediate and deep parts of the aquifer. A vertical-flow log conducted during the TANC study found that approximately 61 percent of the flow to the well is contributed from intermediate depths, whereas 35 percent of the flow is from the deep part of the aquifer.

Tritium and CFCs have been detected in the suppy well, and their concentrations appear to be affected by seasonal groundwater pumpage. Concentrations of ^{3}H and CFCs were highest in samples collected during the spring and summer, when water demand was high, and the lowest concentrations were found during the winter, when water demand was low. ^{14}C also displayed seasonal variability, having the lowest concentrations measured during winter and the highest concentrations measured during spring and summer. The seasonal variability in ^{14}C values reflects a larger component

of young groundwater reaching the well during the high pumping season. In addition, the detection of ^{3}H indicates that bomb-derived ^{14}C likely was present. Because the tracer concentrations indicate a wide range of ages are present in the aquifer, and the single LPMs (PFM, EMM, EPM, PEM, and DM) are incapable of producing tracer concentrations that match measured ^{14}C and ^{3}H concentrations at the well, these results indicate that a binary mixing model could be appropriate for this well.

The interpretation of the groundwater age distribution from the supply well relied on an initial interpretation of the groundwater ages for monitoring-well sites in the vicinity of the well site. The interpretation of age from the monitoring wells helped establish approximate ages of native groundwater at different depths that were unaffected by the movement of young groundwater to deeper depths, the approximate age of young groundwater, and travel times through the thick unsaturated zone. These interpretations were determined from visual inspection of binary mixing (BMM-DM-DM) plots of tracers (primarily ^{3}H, ^{3}He$_{trit}$, and ^{14}C) and the optimization algorithm in TracerLPM. That analysis found the average travel time through the unsaturated zone was about 15 years and the mean age of young groundwater was about 22 years. The details of that portion of the analysis and results are provided in Bexfield and others (2012).

The age distribution of groundwater from the supply well was determined by modeling tracer concentrations by using a binary mixing model composed of two dispersion models (BMM-DM-DM): a dispersion model for the young fraction and a dispersion model for the old fraction. The old fraction that was modeled by the second component in the BMM represents the natural state of the groundwater system prior to the presence of young groundwater at deeper depths due to groundwater pumping. Other models of the second component, such as a BMM-DM-EPM and BMM-DM-PEM, could have been chosen and tested. Because the recharge area (Rio Grande) is far away from the well, the EPM was considered, but the EPM conceptual model is based on a well completed through the entire thickness of an aquifer. Alternatively, the well is configured much like the PEM, but recharge is not areally distributed as in the PEM conceptual model. Both of those models might give acceptable fit results, but were not tested in the original analysis. It should be noted that any of these models would produce results that were non-unique because only ^{14}C data can be used to constrain the age distribution of the old, natural groundwater system The PFM was ruled out because it is known that the well captures old water that is a mixture of ages between 4,000 and 23,000 years.

The model parameters for the old fraction (mean age and DP) were approximated with a DM having mean ages that ranged from 12,000 to 17,200 years and a dispersion parameter of 0.1. If old groundwater has flow path distances on the scale of kilometers to several kilometers, the dispersion parameter would correspond to longitudinal dispersivities that range from about 100 to 1,000 meters, which are reasonable values (Domenico and Schwartz, 1998). The DM that results from this combination of ages and dispersion parameter produces a broad distribution of ages within the old fraction of groundwater. In the optimization of the model, the dispersion parameter of the old fraction was held constant, while the mean age was manually adjusted until the model ^{14}C values matched the measured ^{14}C values after optimization of the young fraction. The DM used for the young fraction

had a dispersion parameter of about 0.01, which simulates tracer behavior somewhat similar to piston-flow. For the optimization of the young fraction, the dispersion parameter was allowed to vary only between 0.01 and 0.03, while the age was allowed to vary only between 20 and 24 years. The unsaturated-zone travel time was 15 years.

For samples collected in June 2007 and May 2009, the mean ages of young and old groundwater in the BMM-DM-DM were determined approximately from graphs of ^{14}C and ^{3}H (fig. 31). These samples plotted along a binary mixing line composed of old water having a mean age of about 12,000 and 13,000 years, and young water having a mean age of about 22 years. These estimates of mean age for the two fractions were then used to constrain the best-fit algorithm on the *TracerTracerFits* worksheet (fig. 32).

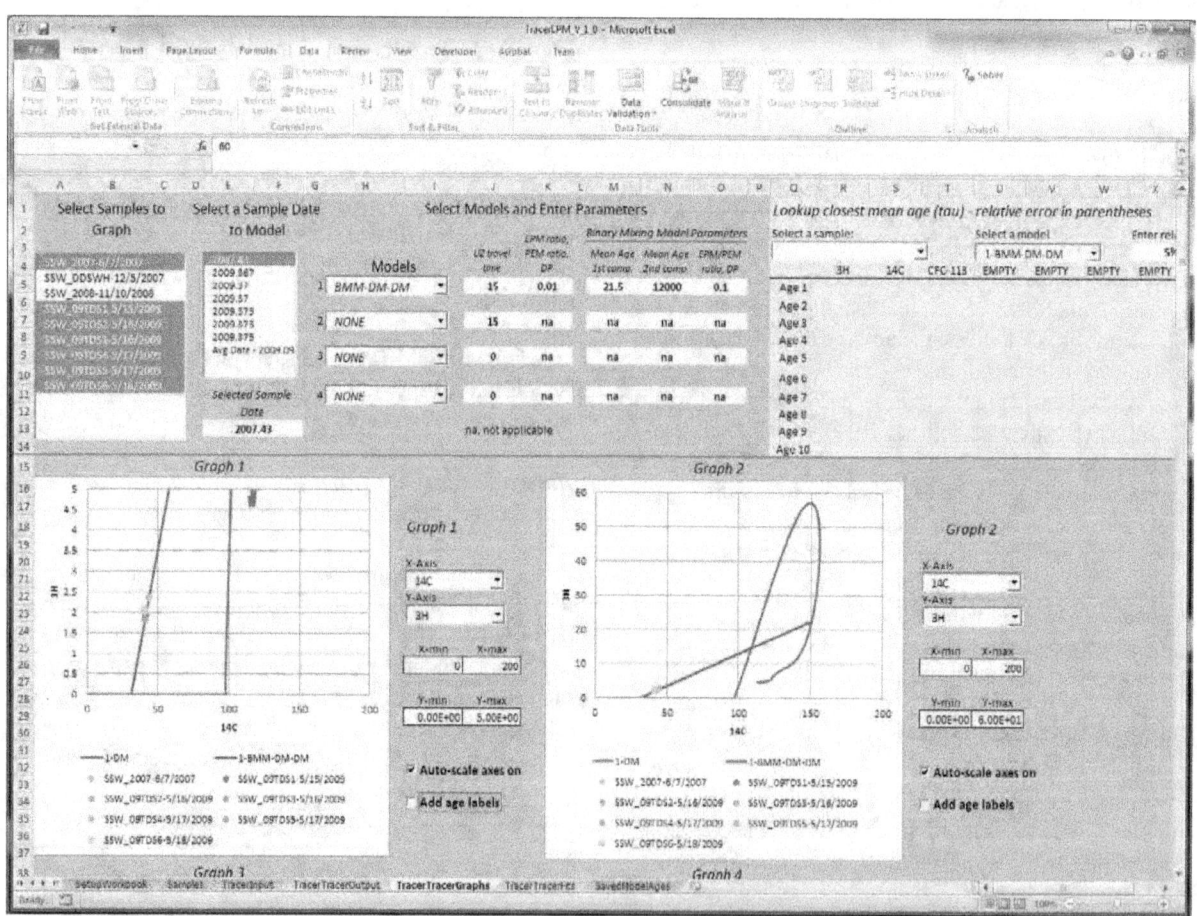

Figure 31.   Screenshot of the *TracerTracerGraph* worksheet used to analyze the mean age and age distribution of groundwater in a public-supply well in Albuquerque, New Mexico. Graphs are of carbon-14 (young) and tritium. Samples were collected from the public-supply well in June 2007 and May 2009 (Bexfield and others, 2012).

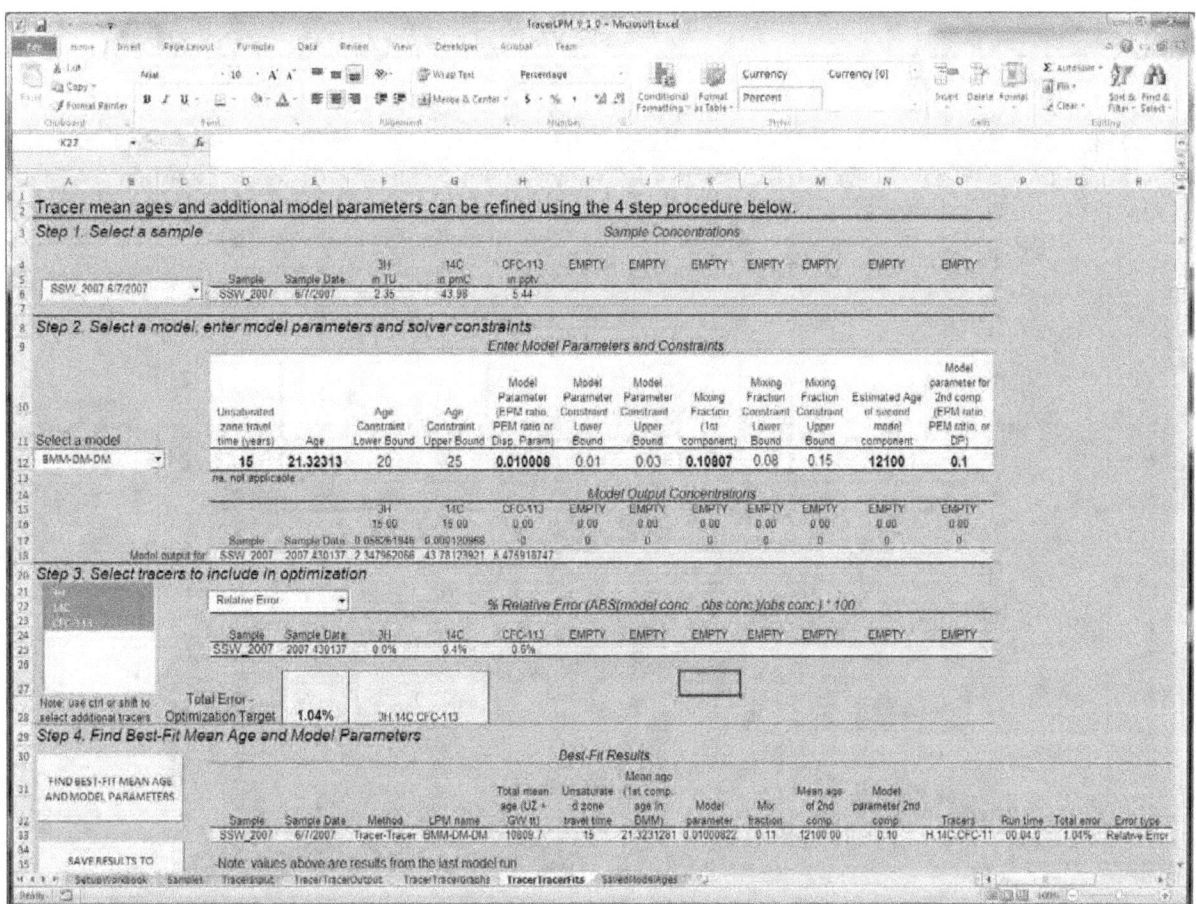

Figure 32. Screenshot of the worksheet *TracerTracerFits* showing the model and constraints used in the best-fit algorithm.

Best-fit results indicate the mean age of young groundwater in the June 2007 sample was 21 years and the sample had a young-water fraction of 10.8 percent. Five samples collected in May 2009 had a mean age of young water of about 22 years and a mean fraction of 11.5 percent young water. CFC-113 was used to constrain the June 2007 sample and 1 sample from the May 2009 samples. Most of the May 2009 CFC-113 concentrations were difficult to model satisfactorily and could be affected by sampling errors, degradation, or sources of contamination (Bexfield and others, 2012).

For samples collected in December 2007 and November 2008, the same approach was used to model the mean age of groundwater. The mean age of the old fraction of groundwater however, was much older than the mean age of old groundwater derived from the spring and summer samples, which were between 11,000 and 12,000 years (fig. 33). On the basis of graphical analysis, the December 2007 sample was assigned a mean age for old groundwater of 17,200 years, and the November 2008 sample was assigned a mean age for old groundwater of 15,000 years. Results of the best-fit analysis indicate the mean age of young groundwater in the

December 2007 and November 2008 samples was about 21 years and the fraction of young groundwater was 3.0 and 5.5 percent, respectively. The December 2007 sample was collected with a lower capacity pump than the one normally used in the supply well. Consequently, this sample likely had a larger fraction of deep, old groundwater than generally is present under normal operation. Therefore, the fraction of young water estimated from the November 2008 sample is likely the most representative of samples collected from the supply well during the fall and winter (or low pumping) seasons.

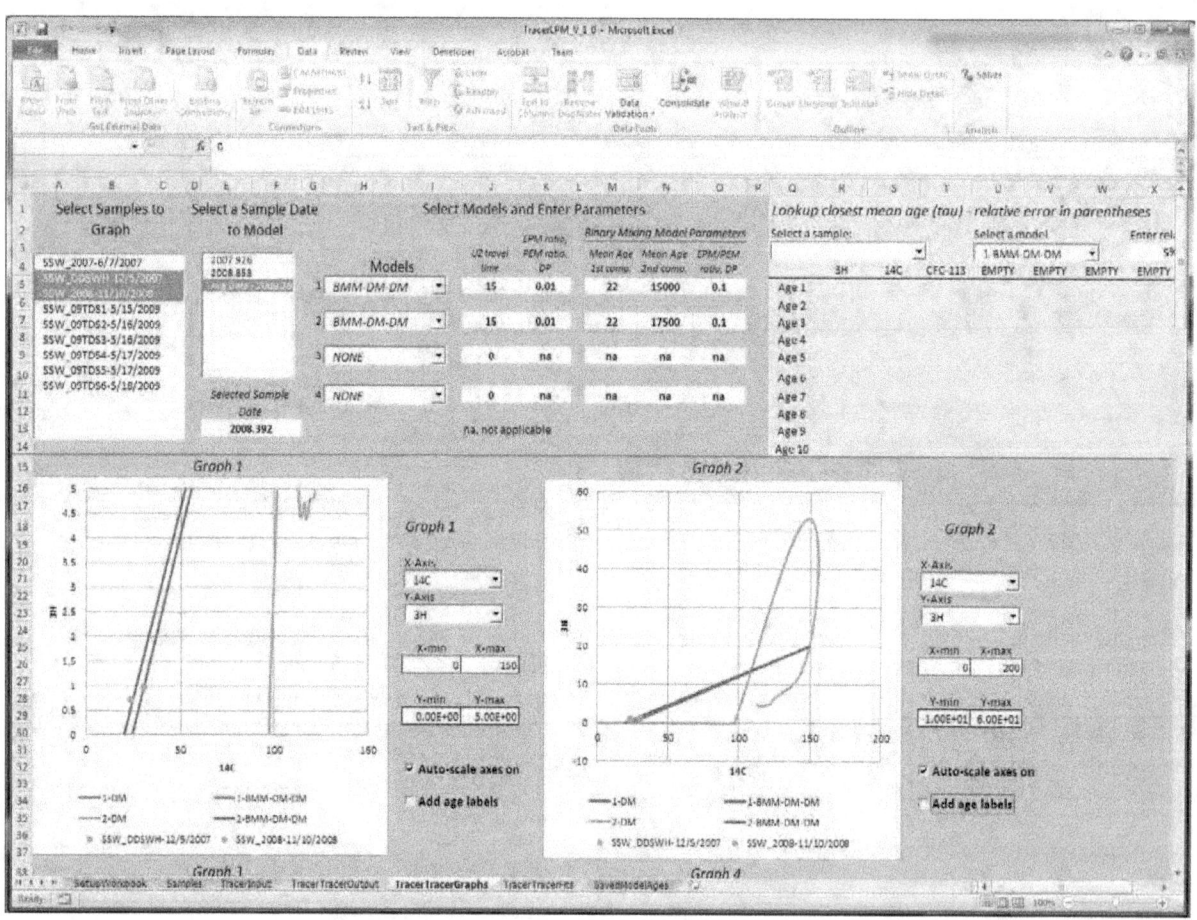

Figure 33. Screenshot of the *TracerTracerGraphs* worksheet showing the models used to determine the mean age and age distribution of samples collected from the public-supply well in December 2007 and November 2008.

The larger fractions of young groundwater and younger ages of old groundwater in public-supply well samples collected during the spring and summer seasons indicate seasonal water demand can have a significant effect on the groundwater quality of local public-supply wells that are screened across large sections of aquifer. Groundwater pumping is highest during summer months when water demand is high. Pumping during the summer months creates prolonged periods of downward vertical gradients and allows shallow, young groundwater to migrate to deeper parts of the aquifer. This indicates that shallow, young groundwater with anthropogenic contaminants is a larger component of groundwater at intermediate depths during the summer than during the winter. Restoration of the natural, upward, vertical gradients during the low pumping season in the fall and winter causes deeper, older groundwater to move upward to shallower parts of the aquifer. Consequently, the age distribution and, ultimately, the vulnerability of the well change with seasonal pumping cycles (fig. 34). These trends also are reflected in concentrations of arsenic, which is associated mostly with the deep aquifer and which increases in concentration in the public-supply well during the fall and winter seasons when water demand is low (Bexfield and others, 2012).

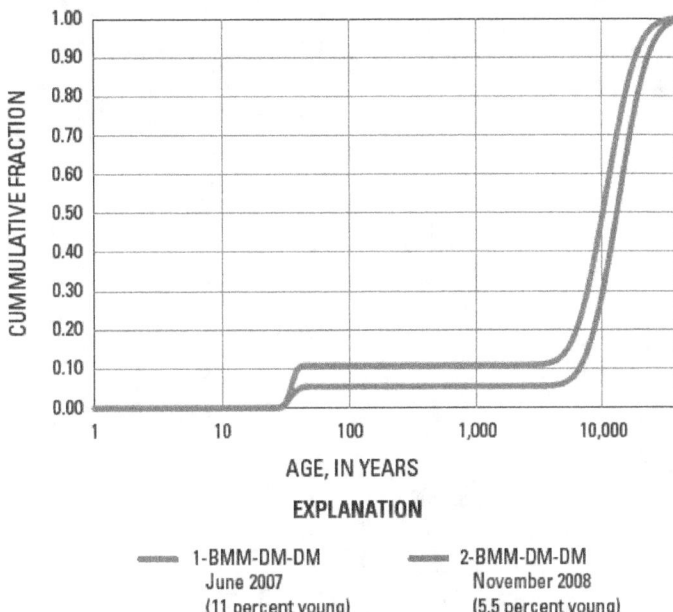

EXPLANATION

━━━ 1-BMM-DM-DM
       June 2007
       (11 percent young)

━━━ 2-BMM-DM-DM
       November 2008
       (5.5 percent young)

Figure 34. Cummulative frequency distribution of age for a public-supply well in Albuquerque, New Mexico. Blue solid line shows frequency distribution of age during high pumping season, and solid red line shows the frequency distribution of age during low pumping season.

Example 3: Residence Time of Water in the Upper Missouri River Basin

Previous work by Michel (2004) showed how the mean residence time of water discharging from the upper Missouri River basin could be estimated from long-term ^{3}H records. Michel (2004) simulated ^{3}H concentrations in the river by using a binary mixing model, where one component represented recent precipitation (prompt outflow less than 1 year old) and the second component represented water derived from the long-term groundwater reservoir of the basin. This example shows how a similar analysis can be done with TracerLPM. Tritium concentrations in the river between 1963 and 1997 (Michel, 2004) were entered into TracerLPM. Tritium in precipitation was estimated by averaging monthly ^{3}H in precipitation records obtained from Michel (updated from 1989) for Lincoln, Nebraska, and Bismarck, North

Dakota (fig. 35). The ^{3}H in precipitation record constructed here had higher peak ^{3}H concentrations than those reported by Michel (2004), although a similar approximation method was used.

As shown in figure 35, ^{3}H concentrations in the Missouri River were approximately one-fourth the ^{3}H concentration in precipitation during the bomb peak years. The decline in ^{3}H in river water was slower than the decline in precipitation, however, and by 1966, ^{3}H concentrations in river water exceed those in precipitation. The water discharged to the river is a mixture of recent precipitation and water from various hydrological systems within the basin; therefore, ^{3}H concentrations in the river represent a mixture of the residence times of recent precipitation (presumably zero or nearly zero) and longer residence times associated with ^{3}H that infiltrated and moved through the subsurface prior to discharging to the river.

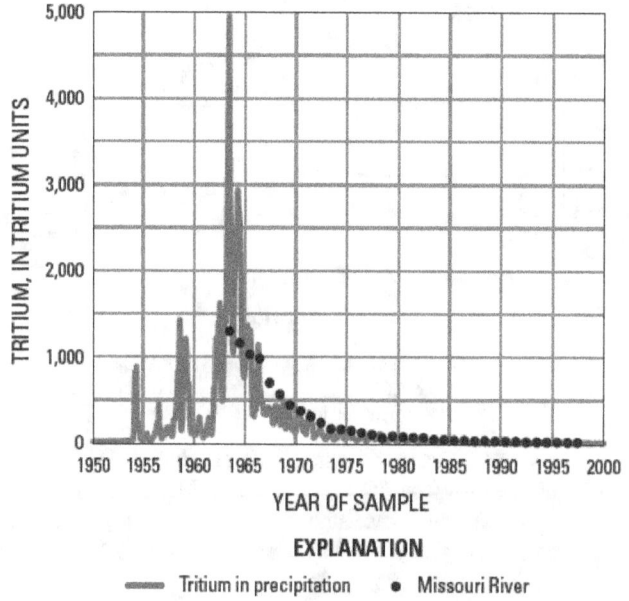

EXPLANATION

——— Tritium in precipitation   ● Missouri River

Figure 35. Comparison of tritium concentrations in precipitation (blue line) and the Missouri River at Nebraska City, Nebraska (black circles). Tritium concentrations in precipitation were estimated by averaging monthly tritium in precipitation records for Lincoln, Nebraska, and Bismarck, North Dakota. Figure modified from Michel (2004).

For the binary mixing model used by Michel (2004), ^{3}H in recent precipitation was simulated with a PFM having a mean age of 0 years, and ^{3}H in the groundwater reservoir was simulated with an EMM. TracerLPM is configured so that only one component of a BMM can be optimized, and this component is the first one given in the model name. This requires that the mean age (and model parameter) of the second-named component is assigned. To accommodate the assumption that the PFM component has a mean age of zero or near zero (instantaneous discharge of the tracer), the model chosen for optimization on the *TimeSeriesFits* worksheet was the BMM-EMM-PFM. The Missouri River model was optimized by choosing constraints on the mean age to vary between 2 and 8 years and the mixing fraction to vary between 0.5 and 1. These constraints yielded a mean age of 4.3 years for the EMM and a mixing fraction (old fraction) of 0.84 (or 84 percent groundwater; fig. 36). Michel's analysis yielded 90 percent groundwater with a mean age of 4 years, which was in close agreement with TracerLPM. Differences

between results was most likely related to small differences between the ^{3}H precipitation records used by Michel and in this analysis, as noted previously, and to TracerLPM's ability to discretize the search domain into a finer resolution than the manual calibration performed by Michel. Tritium concentrations used by Michel had a peak around 4400 TU, whereas the ^{3}H input record used here had a peak concentration around 4900 TU. As a result, concentrations of ^{3}H in old groundwater were likely higher than predicted by Michel. Consequently, a lower contribution of old water was required to match measured ^{3}H in the river. Nonetheless, this example illustrates how TracerLPM can be used in time-series mode, and how some binary mixing models could be useful for watershed residence-time analysis. The BMM-EMM-PFM model could describe situations in which runoff or shallow permeable layers deliver precipitation quickly to streams and are underlain by aquifers with longer residence times that deliver mixtures of older groundwater.

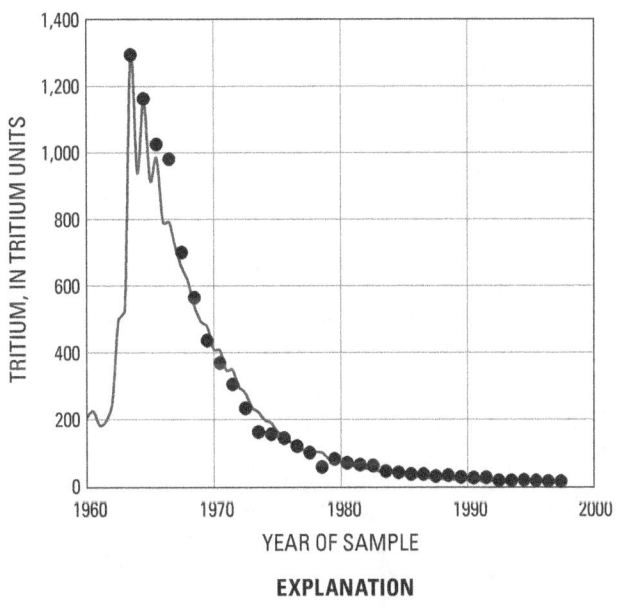

EXPLANATION

—— 1-BMM-EMM-PFM:3H ● Missouri River-7/1/1963: 3H

Figure 36. Comparison of results from the binary mixing model with long-term tritium record for the upper Missouri River. The binary mixing model is composed of an EMM and a PFM: BMM-EMM-PFM. The EMM has a mean age of 4.3 years and makes up 84 percent of the sample, and the PFM has a mean age of 0.

In addition to residence-time analysis, the BMM-EMM-PFM model could be used to hindcast nitrate concentrations in the watershed if the history of nitrate in recharge could be reconstructed. As an example, the model was used to simulate a hypothetical response of nitrate concentrations in the river using the nitrate input history used in example 1 for Modesto, California. The input history for the Missouri River basin is most likely different but presumably has increased over time. In addition, the response was simulated for groundwater in which no denitrification occurred and a second scenario in which a nitrate decay constant of 0.1 per year was applied (fig. 37). The simulations show that in the absence of denitrification, nitrate concentrations in the river largely reflect the nitrate input variations with some delay (lagtime), but when denitrification in groundwater occurs, nitrate concentrations can be significantly attenuated.

In forecasting mode, TracerLPM can be used to explore future responses of the river to different nitrate input scenarios in the watershed. As shown in figure 38, a river basin with BMM-EMM-PFM age distributions in discharge, like that of the Missouri River, can have a rapid partial response, followed by a longer period of gradual response, after a sudden change in nitrate loading on the watershed. The quick initial response in this example is in contrast to the delayed initial response illustrated in the Modesto supply well example, where young components of groundwater were absent.

The results of Michel's (2004) analysis have important implications for ^{3}H and [3]He dating in groundwater. If river water is determined to be the major source of recharge to a groundwater system and the river has long-term ^{3}H records, a similar ^{3}H response function to the one developed here could be created and substituted for the precipitation-derived input function for [3]H. The ^{3}H response function would then be used to evaluate tracer data from groundwater samples (Stute and others, 1997).

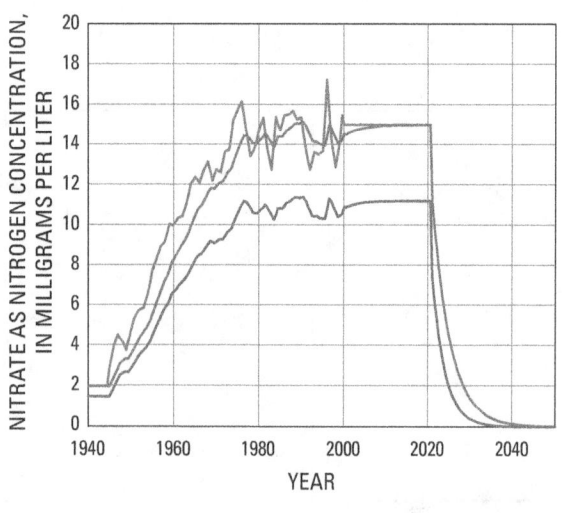

EXPLANATION

— NO$_3$-N input concentration
— Scenario 1: NO$_3$-N output concentration - No decay
— Scenario 2: NO$_3$-N output concentration - 0.1 per year

Figure 37.   Hypothetical results for a simulated response of nitrate concentrations in the Missouri River Basin for a made-up nitrate input history for two different scenarios. Blue line is the nitrate input history in recharge, red line is the nitrate response in the river with no denitrafication, and the purble line is the nitrate response in the river with groundwater having a decay rate of 0.1 per year.

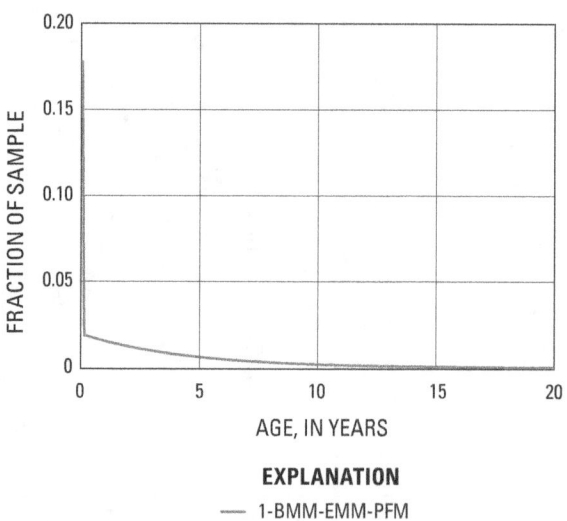

EXPLANATION

— 1-BMM-EMM-PFM

Figure 38.   Modeled age distribution of the upper Missouri River.

Acknowledgments

The funding for this report and development of this program was provided by the U.S. Geological Survey National Water-Quality Assessment program ("Transport of Anthropogenic and Natural Contaminants to supply wells" topical team) and the National Research Program.

Disclaimer

This workbook program was developed by using Microsoft® Visual Basic® for Applications (VBA) code and an Excel add-in written in the C++ programming language using Microsoft Visual Studio 2010 and Microsoft Office Excel 2010 XLL Software Developer Kit. The workbook and add-in are used to interpret tracer data with simple mathematical models of steady-state flow for different hydrogeolgical configurations. All efforts have been made to ensure the program makes accurate determinations; however, it is possible that the code contains errors or that certain uses of the program can give unrealistic or incorrect results. Users are encouraged to keep an original copy of the workbook and notify the author if any errors are found.

References Cited

Aitchison, T.C., Leese, Morven, Michczynska, D.J., Mook, W.G., Otlet, R.L., Ottaway, B.S., Pazdur, M.F., Van Der Plicht, Johannes, Reimer, P.J., Robinson, S.W., Scott, E.M., Stuiver, Minze, and Weninger, Bernhard, 1989, A comparison of methods used for the calibration of radiocarbon dates: Radiocarbon, v. 31, no. 3, p. 846–864.

Amin, I.E., and Campana, M.E., 1996, A general lumped parameter model for the interpretation of tracer data and transit time calculation in hydrologic systems: Journal of Hydrology, v. 179, no. 1–4, p. 1–21.

Andrews, J.N., and Lee, D.J., 1979, Inert gases in groundwater from the Bunter Sandstone of England as indicators of age and paleoclimatic trends: Journal of Hydrology, v. 41, no. 3–4, p. 233–252.

Andrews, J.N., Giles, I.S., Kay, R.L.F., Lee, D.J., Osmond, J.K., Cowart, J.B., Fritz, Peter, Barker, J.F., and Gale, J., 1982, Radioelements, radiogenic helium and age relationships for groundwaters from the granites at Stripa, Sweden: Geochimica et Cosmochimica Acta, v. 46, no. 9, p. 1533–1543.

Appelo, C.A.J, and Postma, Dieke, 1996, Geochemistry, groundwater and pollution (rev. ed.): Rotterdam, Netherlands, CRCPress/Balkema, 536 p.

Bexfield, L.M., Jurgens, B.C., Crilley, D.M., and Christenson, S.C., 2012, Hydrogeology, water chemistry, and transport processes in the zone of contribution of a public-supply well in Albuquerque, New Mexico: U.S. Geological Survey Scientific Investigations Report 11-5182, 114 p.

Böhlke, J.K., 2002, Groundwater recharge and agricultural contamination: Hydrogeology Journal, v. 10, no. 1, p. 153–179.

Böhlke, J.K., 2002, Groundwater recharge and agricultural contamination- Erratum. Hydrogeology Journal, v. 10, p. 438-439.

Böhlke, J.K., 2006, TRACERMODEL1. Excel workbook for calculation and presentation of environmental tracer data for simple groundwater mixtures, pt. 10.3 in app. III: Data evaluation software of International Atomic Energy Agency, ed., Use of chlorofluorocarbons in hydrology: A guidebook: Vienna, Austria, International Atomic Energy Agency Publishing Section, p. 239–246, available online at URL http://www-pub.iaea.org/MTCD/Publications/PDF/Pub1238_web.pdf

Böhlke, J.K., and Denver, J.M., 1995, Combined use of groundwater dating, chemical, and isotopic analyses to resolve the history and fate of nitrate contamination in two agricultural watersheds, Atlantic coastal plain, Maryland: Water Resources Research, v. 31, no. 9, p. 2319–2339.

Böhlke, J.K., and Krantz, D.E., 2003, Isotope geochemistry and chronology of offshore ground water beneath Indian River Bay, Delaware: U.S. Geological Survey Water-Resources Investigations Report 03-4192, 37 p, available online at URL http://pubs.water.usgs.gov/wri03-4192

Brown, C.J., Starn, J.J., Stollenwerk, K.G., Mondazzi, R.A., and Trombley, T.J., 2009, Aquifer chemistry and transport processes in the zone of contribution to a public-supply well in Woodbury, Connecticut, 2002–06: U.S. Geological Survey Scientific Investigations Report 2009-5051, 158 p.

Burden, R.L., and Faires, J.D., 2005, Numerical analysis (8th ed.): Belmont, Calif., Thomson Brooks/Cole, 847 p.

Burow, K.R., Jurgens, B.C., Kauffman, L.J., Phillips, S.P., Dalgish, B.A., and Shelton, J.L., 2008, Simulations of ground-water flow and particle analysis in the zone of contribution of a public-supply well in Modesto, eastern San Joaquin Valley, California: U.S. Geological Survey Scientific Investigations Report 2008-5035, 41 p.

Clark, I.D., and Fritz, Peter, 1997, Environmental isotopes in hydrogeology: Boca Raton, Fla., Lewis Publishers, 328 p.

Cook, P.G., and Böhlke, J.K., 2000, Determining timescales for groundwater flow and solute transport, chap. 1 *in* Cook, P. G., and Herczeg, A. L., eds., Environmental tracers in subsurface hydrology: Boston, Mass., Kluwer Academic Publishers, p. 1–30.

Cook, P.G., and Herczeg, A.L., eds., Environmental tracers in subsurface hydrology: Boston, Mass., Kluwer Academic Publishers, 529 p.

Doney, S.C., Glover, D.M, and Jenkins, W.J., 1992, A model function of the global bomb tritium distribution in precipitation, 1960–1986: Journal of Geophysical Research, v. 97, no. C4, p. 5481–5492.

Domenico, P.A., and Schwartz, F.W., 1998, Physical and chemical hydrogeology (2d ed.): New York, John Wiley and Sons, 506 p.

Eberts, S.M., Erwin, M.L., and Hamilton, P.A., 2005, Assessing the vulnerability of public-supply wells to contamination from urban, agricultural, and natural sources: U.S. Geological Survey Fact Sheet 2005-3022, 4 p.

Eberts, S.M., Böhlke, J.K., Kauffmann, L.J, and Jurgens, B.C., 2012, Comparison of particle-tracking and lumped-parameter age-distribution models for evaluating vulnerability of production wells to contamination: Hydrogeolgy Journal, v. 20, no. 1, p. 1–20, available online at URL http://www.springerlink.com/content/6177t487m60545n1/fulltext.pdf

El-Kadi, Aly I., Plummer, L.N., and Aggarwal, Pradeep, 2011, NETPATH-WIN: An interactive user version of the mass-balance model, NETPATH: Ground Water, v. 49, no. 4, p. 593–599, available online at URL http://onlinelibrary.wiley.com/doi/10.1111/j.1745-6584.2010.00779.x/pdf

Eriksson, Erik, 1971, Compartment models and reservoir theory: Annual Review of Ecology and Systematics, v. 2, p. 67–84, available online at URL http://www.annualreviews.org/doi/pdf/10.1146/annurev.es.02.110171.000435

Focazio, M.J., Plummer, L.N., Böhlke, J.K., Busenberg, Eurybiades, Bachman, L.J., and Powers, D.S., 1998, Preliminary estimates of residence times and apparent ages of ground water in the Chesapeake Bay watershed, and water-quality data from a survey of springs: U.S. Geological Survey Water-Resources Investigations Report 97-4225, 75 p.

Fylstra, Daniel, Lasdon, Leon, Watson, John, and Waren, Allan, 1998, Design and use of the Microsoft Excel Solver: Interfaces, v. 28, no. 5, p. 29–55.

Hua, Quan, and Barbetti, Mike, 2004, Review of tropospheric bomb ^{14}C data for carbon cycle modeling and age calibration purposes: Radiocarbon, v. 46, no. 3, p. 1273–1298.

International Atomic Energy Agency, ed., 2006, Use of chlorofluorocarbons in hydrology: A guidebook: Vienna, Austria, International Atomic Energy Agency Publishing Section, 291 p., available online at URL http://www-pub.iaea.org/MTCD/Publications/PDF/Pub1238_web.pdf

Jurgens, B.C., Burow, K.R., Dalgish, B.A., and Shelton, J.L., 2008, Hydrogeology, water chemistry, and factors affecting the transport of contaminants in the zone of contribution of a public-supply well in Modesto, eastern San Joaquin Valley, California: U.S. Geological Survey Scientific Investigations Report 2008-5156, 78 p.

Katz, B.G., Böhlke, J.K., and Hornsby, H.D., 2001, Timescales for nitrate contamination of spring waters, northern Florida, USA, chap. 11 *in* Alley, W. M., ed., Regional ground-water quality: New York, Van Nostrand Reinhold, p. 255–294.

Katz, B.G., Hornsby, H.D., Böhlke, J.K., and Mokray, M.F., 1999, Sources and chronology of nitrate contamination in spring waters, Suwanee River basin, Florida: U.S. Geological Survey Water Resources Investigations Report 99-4252, 54 p.

Katz, B.G., McBride, W.S., Hunt, A.G., Crandall, C.A., Metz, P.A., Eberts, S.M., and Berndt, M.P., 2008, Vulnerability of a public supply well in a karstic aquifer to contamination: Ground Water, v. 47, no. 3, May–June 2009, p. 438–452.

Kazemi, G. A., Lehr, J. H., and Perrochet, Pierre, 2006, Modeling of groundwater age and residence-time distributions, chap. 6 *in* Groundwater Age: Hoboken, NJ, John Wiley & Sons, p. 204–253.

Kinzelbach, Wolfgang; Aeschbach, Werner; Alberich, Carmen; Goni, I.B.; Beyerle, Urs; Brunner, Philip; Chiang, W.-H.; Rueedi, Joerg; and Zoellmann, Kai, 2002, A survey of methods for groundwater recharge in Arid and Semi-arid regions: Early Warning and Assessment Report Series: Nairobi, Kenya, Division of Early Warning and Assessment, United Nations Environment Programme, 101 p., available online at URL http://www.unep.org/PDF/groundwaterrecharge.pdf

Kreft, Andrzej, and Zuber, Andrzej, 1978, On the physical meaning of the dispersion equation and its solution for different initial and boundary conditions: Chemical Engineering Science, v. 33, no. 11, p. 1471–1480, available online at URL http://www.sciencedirect.com/science/article/pii/0009250978851963

Landon, M.K., Clark, B.R., McMahon, P.B., McGuire, V.L., and Turco, M.J., 2008, Hydrogeology, Chemical-characteristics and transport processes in the zone of contribution of a public-supply well in York, Nebraska: U.S. Geological Survey Scientific Investigations Report 2008-5050, 149 p.

Lindsey, B.D., Phillips, S.W., Donnelly, C.A., Speiran, G.K., Plummer, L.N., Böhlke, J.K, Focazio, M.J., Burton, W.C., and Busenberg, Eurybiades, eds., 2003, Residence times and nitrate transport in ground water discharging to streams in the Chesapeake Bay Watershed: U.S. Geological Survey Water-Resources Investigations Report 03-4035, 201 p., available online at URL http://pa.water.usgs.gov/reports/wrir03-4035.pdf

Long, A.L., and Putnam, L.D., 2006, Translating CFC-based piston ages into probability density functions of ground-water age in karst: Journal of Hydrology, v. 330, no. 3–4, p. 735–747.

Maloszewski, Piotr, and Zuber, Andrzej, 1982, Determining the turnover time of groundwater systems with the aid of environmental tracers: 1. Models and their applicability: Journal of Hydrology, v. 57, no. 3–4, p. 207–231.

Maloszewski, P., Rauert, W., Stichler, W., and Herrmann, A., 1983, Application of flow models to an Alpine catchment area using tritium and deuterium data: Journal of Hydrology, v. 66, p. 319-330.

Maloszewski, Piotr, and Zuber, Andrzej, 1996, Lumped parameter models for the interpretation of environmental tracer data, chap. 2 in International Atomic Energy Agency, ed., Manual on mathematical models in isotope hydrogeology, TECDOC-910: Vienna, Austria, International Atomic Energy Agency Publishing Section, p. 9–58.

McCormac, F.G., Hogg, A.G., Blackwell, P.G., Buck, C.E., Higham, T.F.G., and Reimer, P.J., 2004, SHCal04 Southern Hemisphere calibration, 0–11.0 cal kyr BP: Radiocarbon, v. 46, no. 3, p. 1087–1092.

McCormac, F.G., Hogg, A.G., Higham, T.F.G., Lynch-Stieglitz, Jean, Broecker, W.S., Baillie, M.G.L., Palmer, J., Xiong, L., Pilcher, J.R., Brown, D., and Hoper, S.T., 1998, Temporal variation in the interhemispheric ^{14}C offset: Geophysical Research Letters, v. 25, no. 9, p. 1321–1324.

McMahon, P.B., Böhlke, J.K., Kaufmann, L.J., Kipp, K.L., Landon, M.K., Crandall, C.A., Burow, K.R., and Brown, C.J., 2008a, Source and transport controls on the movement of nitrate to public supply wells in selected principal aquifers of the United States: Water Resources Research, v. 44, 17 p., available online at URL http://www.agu.org/pubs/crossref/2008/2007WR006252.shtml

McMahon, P.B., Burow, K.R., Kaufmann, L.J., Eberts, S.M., Böhlke, J.K., and Gurdak, J.J., 2008b, Simulated response of water quality in public supply wells to land use change: Water Resources Research, v. 44, 16 p., available online at URL http://www.agu.org/pubs/crossref/2008/2007WR006731.shtml

Michel, R.M., 1989, Tritium deposition in the Continental United States, 1953–1983: U.S. Geological Survey Water-Resources Investigations Report 89-4072, 46 p.

Michel, R.M., 2004, Tritium hydrology of the Mississippi River Basin: Hydrological Processes, v. 18, no. 7, p. 1255–1269.

Mook, W.G., and Van Der Plicht, Johannes, 1999, Reporting ^{14}C activities and concentrations: Radiocarbon, v. 41, no. 3, p. 227–239.

Ozyurt, N.N., and Bayari, C.S., 2005, LUMPED unsteady: a Visual Basic code of unsteady-state lumped-parameter models for mean residence time analyses of groundwater systems: Computers & Geosciences, v. 31, no. 3, p. 329–341.

Ozyurt, N.N., and Bayari, C.S., 2003, LUMPED: a Visual Basic code of lumped-parameter models for mean residence time analyses of groundwater systems: Computers & Geosciences, v. 29, no. 1, p. 79–90.

Parkhurst, D.L., and Appelo, C.A.J., 1999, User's guide to PHREEQC (ver. 2)—A computer program for speciation, batch-reaction, one-dimensional transport, and inverse geochemical calculations: U.S. Geological Survey Water-Resources Investigations Report 99-4259, 310 p.

Parkhurst, D.L., and Charlton, S.R., 2008, NetpathXL—An Excel interface to the program NETPATH: U.S. Geological Survey Techniques and Methods 6-A26, 11 p.

Phillips, J.D., Duval, J.S., and Ambroziak, R.A., 1993, National geophysical data grids; gamma-ray, gravity, magnetic, and topographic data for the conterminous United States: U.S. Geological Survey Digital Data Series DDS-9, accessed August 30, 2010 at URL http://crustal.usgs.gov/geophysics/North_America.html

Plummer, L.N., Prestemon, E.C., and Parkhurst, D.L., 1994, An interactive code (NETPATH) for modeling NET geochemical reactions along a flow PATH, version 2.0: U.S. Geological Survey Water-Resources Investigations Report 94-4169, 130 p., accessed June 7, 2010, at URL http://pubs.er.usgs.gov/usgspubs/wri/wri944169

Plummer, L.N., Busenberg, Eurybiades, Böhlke, J.K., Nelms, D.L., Michel, R.L., and Schlosser, Peter, 2001, Groundwater residence times in Shenandoah National Park, Blue Ridge Mountains, Virginia, USA: a multi-tracer approach: Chemical Geology, v. 179, no. 1–4, p. 93–111.

Plummer, L.N., Bexfield, L.M., Anderholm, S.K., Sanford, W. E., and Busenberg, Eurybiades, 2004, Geochemical characterization of groundwater flow in the Santa Fe Group aquifer system, Middle Rio Grande Basin, New Mexico: U.S. Geological Survey Water-Resources Investigations Report 03-4131, 395 p.

Reed, J.C., and Bush, C.A., 2005, Generalized geologic map of the United States, Puerto Rico, and the U.S. Virgin Islands, U.S. Geological Survey, available online at URL http://pubs.usgs.gov/atlas/geologic/

Reimer, P.J., Brown, T.A., and Reimer, R.W., 2004, Discussion: Reporting and calibration of post-bomb ^{14}C data: Radiocarbon, v. 46, no. 3, p. 1299–1304.

Reimer, P.J.; Baillie, M.G.L.; Bard, Edouard; Bayliss, Alex; Beck, J.W.; Blackwell, P.G.; Bronk Ramsey, Christopher; Buck, C.E.; Burr, G.S.; Edwards, R.L.; Friedrich, Michael; Grootes, P.M.; Guilderson, T.P.; Hajdas, Irka; Heaton, T.J.; Hogg, A.G.; Hughen, K.A.; Kaiser, K.F.; Kromer, B.; McCormac, F.G.; Manning, S.W.; Reimer, R.W.; Richards, D.A.; Southon, J.R.; Talamo, Sahra; Turney, C.S.M.; Van Der Plicht, Johannes; and Weyhenmeyer, C.E., 2009, IntCal09 and Marine09 radiocarbon age calibration curves, 0–50,000 years cal BP: Radiocarbon, v. 51., no. 4, p. 1111–1150.

Schlosser, Peter; Stute, Martin; Sonntag, Christian; and Munnich, K.O., 1989, Tritiogenic [3]He in shallow groundwater: Earth and Planetary Science Letters, v. 94, no. 3–4, p. 245–256.

Solomon, D.K., Hunt, A., and Poreda, R. ., 1996, Source of radiogenic helium 4 in shallow aquifers: Implications for dating young groundwater: Water Resources Research, v. 32, no. 6, p. 1805–1813.

Stuiver, Minze, 1961, Variations in radiocarbon concentration and sunspot activity: Journal of Geophysical Research, v. 66, no. 1, p. 273–276.

Stuiver, Minze, 1965, Carbon-14 content of the 18th- and 19th-century wood: variations correlated with sunspot activity: Science, v. 149, no. 3683, p. 533–535.

Stuiver, Minze, 1982, A high-precision calibration of the AD radiocarbon time scale: Radiocarbon, v. 24, no. 1, p. 1–26.

Stuiver, Minze, and Pollach, H. A., 1977, Discussion—Reporting of ^{14}C data: Radiocarbon, v. 19, no. 3, p. 353–363.

Stute, M., Deak, J., Revesz, K., Böhlke, J.K., Deseo, E., Weppernig, R., and Schlosser, P., 1997, Tritium/3He dating of river infiltration: An example from the Danube in the Szigetköz area, Hungary: Ground Water, v. 35, p. 905-911.

Torgensen, Thomas, and Clarke, W.B., 1985, Helium accumulation in groundwater, I: An evaluation of sources and the continental flux of crustal [4]He in the Great Artesian Basin, Australia: Geochimica et Cosmochimica Acta, v. 49, no. 5, p. 1211–1218.

Vogel, J.C., 1967, Investigation of groundwater flow with radiocarbon: Isotopes in hydrology, proceedings from Conference on Isotopes in Hydrology, Vienna, International Atomic Energy Agency, p. 355–368.

Weissmann, G.S., Zhang, Y., LaBolle, E.M., and Fogg, G.E., 2002, Dispersion of groundwater age in an alluvial aquifer system: Water Resources Research, v. 38, no. 10, p. 16-1–16-13.

Zoellmann, Kai; Aeschbach-Hertig, Werner; and Beyerle, Urs, 2002, BOXMODEL: An Excel-workbook for the interpretation of transient tracer data (^{3}H, [3]He, CFCs, [85]Kr) , chap. 5 in Kinzelbach, Wolfgang; Aeschbach-Hertig, Werner; Alberich, Carmen; Goni, I.B.; Beyerle, Urs; Brunner, Philip; Chiang, W. -H.; Rueedi, Joerg; and Zoellmann, Kai, A survey of methods for groundwater recharge in Arid and Semi-arid regions: Early Warning and Assessment Report Series: Nairobi, Kenya, Division of Early Warning and Assessment, United Nations Environment Programme, p. 33–43.

Zuber, Andrzej, and Maloszewski, Piotr, 2001, Lumped parameter models, chap. 2 of Mook, W. G., and Yurtsever, Yuecel, eds., Volume 6: Modelling in Environmental isotopes in the hydrological cycle: Principles and applications: Paris, France, UNESCO, Technical Documents in Hydrology, v. 39, no. 1, p. 5–35.

Appendix A: PEM and EPM Derivations

In this appendix, derivations of the PEM and EPM are provided. The derivations follow the mathematical conventions of Kazemi and others (2006); although some variables have been renamed. Both models assume horizontal velocities are constant over the depth of the aquifer and that the darcy flux at any location is equal to the amount of effective infiltration over the area upstream of that location. This implies the well or spring captures water that is representative of the natural age gradient..

Partial Exponential Mixing Model (PEM)

The partial exponential mixing model (PEM) presented in this report is used to describe tracer concentrations from wells that are partially screened in aquifers that can be described by the exponential mixing model (fig. A1). The exponential mixing model (EMM) was derived for a homogeneous aquifer receiving uniform recharge (Vogel, 1967) and frequently has been used to describe tracer concentrations at a well or spring. In this model, the distribution of age in the aquifer increases logarithmically from zero at the water table to infinity at the base of the aquifer. The EMM, as typically defined in past research, requires the well be screened over the entire depth of aquifer; however, in practice this requirement is not frequently met because different well types often have different construction characteristics. Public-supply wells or production wells typically have long-screens, but often are not screened up to the water table and also might not extend to the base of the aquifer. Domestic wells can be screened near the water table but not extend to the base of the aquifer and can have relatively short screens. Monitoring wells often have short screens (less than 10 meters) and can be screened at various depths in an aquifer. Therefore, tracer concentrations from these wells might not be expected to follow outlet tracer concentrations from the traditional EMM, but they could follow a partial exponential model that accounts for the portion of aquifer that is sampled by the well (fig. A1). The formulation of the PEM allows inclusion of well construction information to help validate or test the model.

A homogeneous aquifer with constant thickness (H), and porosity (ϕ), and uniform recharge rate (r), has the following depth-dependent age relation:

$$t(z) = -\frac{H\phi}{r}\ln\left(\frac{H-z}{H}\right) \quad (A1)$$

where, z is the depth below the top of the saturated interval or water table. This equation can be used to determine the mean age of groundwater in the aquifer or any continuous portion of the aquifer by calculating the first moment of any sub-domain of the aquifer:

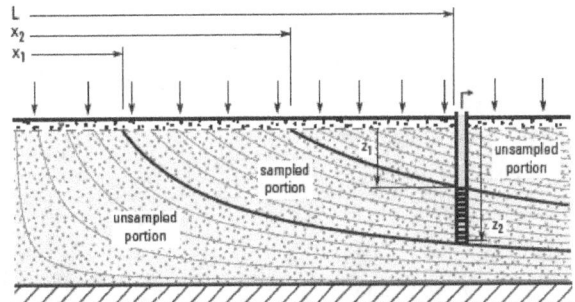

Figure A1. Schematic diagram of idealized aquifer configuration described by the partial exponential mixing model (PEM).

$$\tau_s = -\frac{1}{(z_2 - z_1)}\int_{z_1}^{z_2} \frac{H\phi}{r}\ln\left(\frac{H-z}{H}\right)dz \quad (A2)$$

The general solution to A2 follows:

$$\tau_s = \left(\frac{H}{z_2 - z_1}\right)\left(\frac{H\phi}{r}\right)\left[\begin{array}{c}\left(\frac{H-z_2}{H}\right)\ln\left(\frac{H-z_2}{H}\right) \\ -\left(\frac{H-z_2}{H}\right)-\left(\frac{H-z_1}{H}\right) \\ \ln\left(\frac{H-z_1}{H}\right)+\left(\frac{H-z_1}{H}\right)\end{array}\right] \quad (A3)$$

From this equation, the mean age of the aquifer is found when $z_1 = 0$ and $z_2 = H$:

$$\tau_s = \tau_{aq} = \left(\frac{H\phi}{r}\right)$$

Then, substitute:

$$n_1 = \frac{H}{H-z_1} \text{ and } n_2 = \frac{H}{H-z_2}, \text{ and } \tau_{aq} = \frac{H\phi}{r}$$

$$\tau_s = \left(\frac{1}{\dfrac{1}{n_1}-\dfrac{1}{n_2}}\right)\tau_{aq}\left[\frac{1}{n_1}\ln(n_1)+\frac{1}{n_1}-\frac{1}{n_2}\ln(n_2)-\frac{1}{n_1}\right] \quad (A4)$$

Equation A4 can also be found by calculating the first moment in the x direction, by using the following distance-dependent travel-time equation and equivalent definitions of n_1 and n_2 in the x-direction:

$$t(x) = -\frac{H\phi}{r}\ln\left(\frac{L}{x}\right) \tag{A5}$$

$$n_1 = \frac{L}{x_2} \text{ and } n_2 = \frac{L}{x_1}$$

where

$x_2, x_1,$ and L are horizontal distances from a no-flow boundary (groundwater divide).

The relation between the mean age of groundwater in any sampled portion of the aquifer, τ_s, and the mean age of groundwater in the aquifer, τ_{aq}, allows the direct calculation of outlet tracer concentrations at a well by computing the convolution over the sampled portion. The exit-age frequency distribution of the PEM, $g(t)$, is the same as the EMM, but the convolution is calculated over the sub-domain of ages captured by the well and normalized to 1:

$$\text{PEM}_{g(t)} = \left(\frac{1}{\dfrac{1}{n_1} - \dfrac{1}{n_2}}\right)\frac{1}{\tau_{aq}}exp\left(\frac{-t}{\tau_{aq}}\right); \text{for } t_1 \leq t \leq t_2 \tag{A6}$$

where

t_1 is $\tau_{aq}\ln(n_1)$, and

t_2 is $\tau_{aq}\ln(n_2)$.

Equation A6 is the general solution to the partial exponential mixing model or PEM. This model has three parameters: mean age of the sampled portion τ_s, n_1, and n_2.

Equation A3 can be used to determine the mean age for special cases of z_1 and z_2 (table 1, fig. A2). Case 1 corresponds to the solution of the EMM, where the entire aquifer is sampled by a fully penetrating well. The second case corresponds to a well that is screened from the water table to some arbitrary depth, z_2. The third case corresponds to a situation where the well screen begins at some arbitrary depth, z_1, and extends to the base of the aquifer.

Table 1.   Formulas for mean age of special cases of the partial exponential mixing model (PEM).

Case	z_1	z_2	n_1	n_2	Mean age of sampled portion, τ_s
1	0	H	1	∞	τ_{aq} or aquifer turnover time
2	0	z_2	1	>1	$\left(1+\dfrac{1}{n_2}\right)(\tau_{aq})\left[1-\dfrac{1}{n_2}\ln(n_2)-\dfrac{1}{n_2}\right]$
3	z_1	H	>1	∞	$n_1(\tau_{aq})\left[\dfrac{1}{n_1}\ln(n_1)+\dfrac{1}{n_1}\right] = \tau_{aq}\left[\ln(n_1)+1\right]$

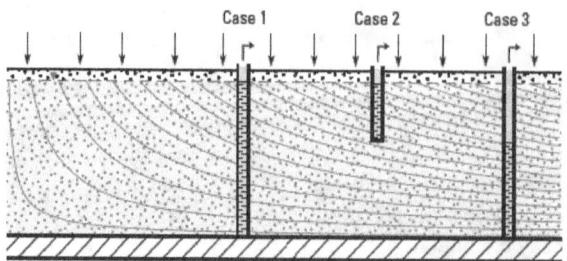

Figure A2.   Special cases of the partial exponential mixing model (PEM) described in table 1.

In the initial version of TracerLPM, only case 3 was implemented, but subsequent versions of TracerLPM will have the full 3-parameter version of the PEM described here. The age distribution of the PEM for case 3 is similar to the EPM. The parameter is related to the same parameter in the exponential piston-flow model by this equation:

$$n_{EPM} = \ln\left(n_{PEM}\right) + 1; \text{when } \tau_s = \tau_{aq} \text{ of the EPM} \qquad \text{(A7)}$$

This relation indicates the difference between the parameters $nEPM$ and $nPEM$ becomes greater as the piston-flow component becomes more dominant in the EPM model, or the top of the well screen becomes closer to the bottom of the aquifer in the PEM.

Exponential Piston-Flow Model

The exponential piston-flow model (EPM) can be derived by combining the travel times of two aquifers, an unconfined and a confined, connected in series (fig. A3). The travel time of a water parcel traveling from the recharge area of the unconfined portion to an outlet point, a well, in the confined portion can be expressed as a function of distance from a no-flow boundary:

$$T\left(x\right) = \int_x^{x^*} \frac{dx}{vx} + \int_{x^*}^{L} \frac{dx}{vx} \qquad \text{(A8)}$$

The age profile in the unconfined portion of the aquifer is logarithmic, with ages ranging from zero at the water table to infinity at the base of the aquifer. Water in the confined portion ages linearly with distance from the unconfined area at x^* to the well at L.

$$T\left(x\right) = \frac{H\phi}{r}\left[\ln\left(\frac{x^*}{x}\right) + \frac{L-x^*}{x^*}\right] = \frac{H\phi}{r}\left[\ln\left(\frac{x^*}{x}\right) + \frac{L}{x^*} - 1\right] \text{ (A9)}$$

The relation between depth, z, and x can be found for the unconfined part of the aquifer by relating the total flux at x^* over the aquifer thickness, H, and the total recharge over x^*–x (fig. A3). These two components must be equal to satisfy the mass balance of the system.

$$q\left(x^*\right)z = \frac{rx^*}{H}z = r\left(x^* - x\right) \qquad \text{(A10)}$$

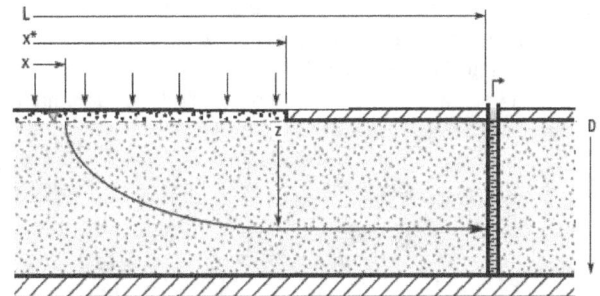

Figure A3. Schematic diagram of idealized aquifer configuration described by the exponential piston-flow model (EPM).

Therefore,

$$x = \left(1 - \frac{z}{H}\right)x^* \qquad \text{(A11)}$$

Equation A9 can be restated in terms of depth:

$$t\left(z\right) = -\frac{H\phi}{r}\left[\ln\left(1 - \frac{z}{H}\right) - \frac{L}{x^*} + 1\right] \qquad \text{(A12)}$$

The mean age of the unconfined portion is the aquifer volume divided by the flow rate:

$$\tau_{x^*} = \frac{V}{Q} = \frac{Hx^*\phi}{rx^*} = \frac{H\phi}{r} \qquad \text{(A13)}$$

For the entire aquifer, the mean age follows:

$$\tau_{aq} = \frac{V_{aq}}{Q_{aq}} = \frac{LH\phi}{rx^*} = \frac{L}{x^*}\tau_{x^*}; \tau_{x^*} = \frac{x^*}{L}\tau_{aq} \qquad \text{(A14)}$$

So, depth as a function of t expressed in terms of the aquifer turnover time follows:

$$Z\left(t\right) = H\left(1 - exp\left(\frac{-Lt}{x^*\tau_{aq}} + \frac{L}{x^*} - 1\right)\right) \qquad \text{(A15)}$$

In order to calculate the exit age distribution at the well, the volumetric flow rate, calculated for the entire aquifer thickness, is needed. At the well, the Darcy flux is (assuming horizontal velocity is constant with depth),and the cumulative discharge with respect to depth is $Q(H) = qH$:

$$Q(t) = qz(t) \tag{A16}$$
$$= rx^* \left(1 - exp \left(\frac{-Lt}{x^* \tau_{aq}} + \frac{L}{x^*} - 1 \right) \right), \ Q(\infty)$$
$$= Q_{aq} = rx^*$$

The exit age distribution is then found by taking the time derivative of the above equation normalized to the total discharge rate:

$$g(t) = \frac{1}{Q_{aq}} \frac{\partial Q(t)}{\partial t} = \frac{L}{x^* \tau_{aq}} exp \left(\frac{-Lt}{x^* \tau_{aq}} + \frac{L}{x^*} - 1 \right) \tag{A17}$$
$$\text{for } t \geq \tau_{aq} \left(1 - \frac{x^*}{L} \right), \ 0$$

Substituting n for $\dfrac{L}{x*}$, the formulation of the EPM reported by Maloszewski and Zuber (1982) is obtained:

$$g(t) = \frac{n}{\tau_{aq}} e^{\left(-\frac{nt}{\tau_{aq}} + n - 1 \right)}, \text{ for } t \geq \tau_{aq} \left(1 - \frac{1}{n} \right), \ 0 \tag{A18}$$

It should be noted that the EPM also could be configured or parameterized like the PEM to account for capture of partial portions of aquifer. This would lead to a model with up to four parameters, however, which would be more difficult to calibrate with tracer data from a single well.

Appendix B: Installation Notes

Compatability

The TracerLPM workbook was developed by using Microsoft Visual Basic® for Applications and the Excel® add-ins, TracerLPMfunctions_32_v_1.xll and TracerLPMfunctions_64_v_1.xll, which are distributed with the workbook, and were written in the C++ programming language by using Microsoft Visual Studio 2010 (version 10) and Microsoft Office Excel 2010 XLL Software Developer Kit.

The TracerLPM workbook and the Excel® add-ins were designed to work with Microsoft® Excel® 2007 versions and later, running on Windows XP and later operating systems. The 32-bit version of the add-in, TracerLPMfunctions_32_v_1.xll, will work with Excel 2007 and the 32-bit version of Excel 2010. The 64-bit version of the add-in TracerLPMfunctions_64_v_1.xll, will work with the 64-bit version of Excel 2010. Users of Excel 2010 should check their version of Excel before downloading and installing the program. To check the version of Excel 2010, select the file menu above the ribbon and select "Help" on the left-hand side of the file menu (fig. B1).

Figure B1.   Screenshot of the *Help* menu selected from the File tab on Excel 2010.

Enabling Macros

This workbook contains Microsoft® Visual Basic® for Applications code — often referred to as "Macros." If the Macro security settings within Excel are set to disable this content, the program will not operate when opened and a security warning will appear between the ribbon and the worksheet (fig. B2).

To change the setting, click the "Options…" button next to the security warning (fig. B2), and select the "Enable this content" radio dial (fig. B3) from the Microsoft® Office Security Options dialog box.

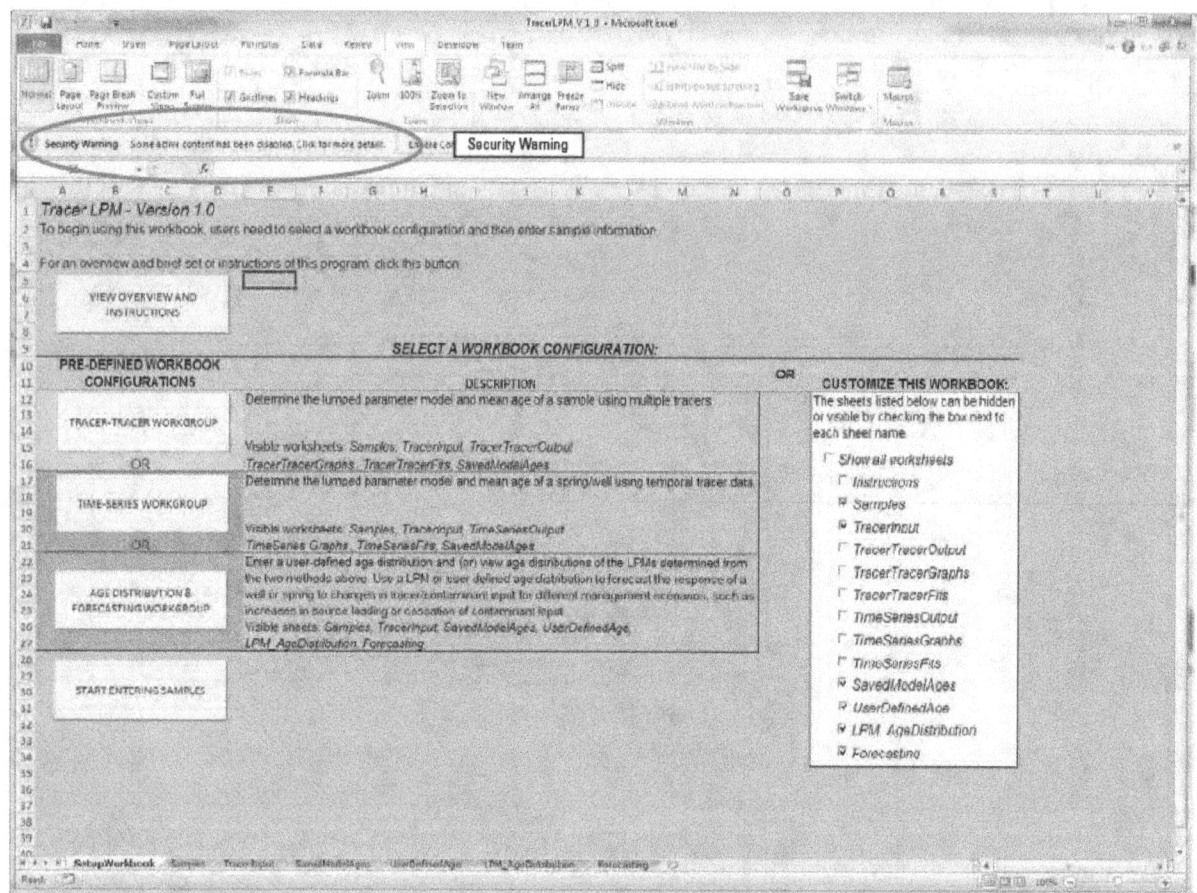

Figure B2. Screenshot of security warning displayed by Excel® when opening the *TracerLPM* workbook when macros are not enabled.

Figure B3. Microsoft Office Security Options dialog box.

Installing the TracerLPM Add-In

TracerLPM is distributed with an Excel® add-in, TracerLPMfunctions.xll. The Windows Installer will automatically copy the Excel add-in to the correct directory. If the installation fails, however, the add-in can be manually installed by copying the TracerLPMfunctions.xll add-in to the same folder as the TracerLPM workbook; the workbook will automatically register and load the add-in when it is opened. If the TracerLPMfunctions.xll add-in is located in another folder, the user will have to manually add the add-in through the Excel Options dialog menu (fig. B4).

To manually add the TracerLPMfunctions.xll add-in, click the "Browse…" button on the Add-Ins dialog box (fig. B5) and navigate to the folder where the TracerLPMfunctions.xll is located. Excel will automatically register the XLL.

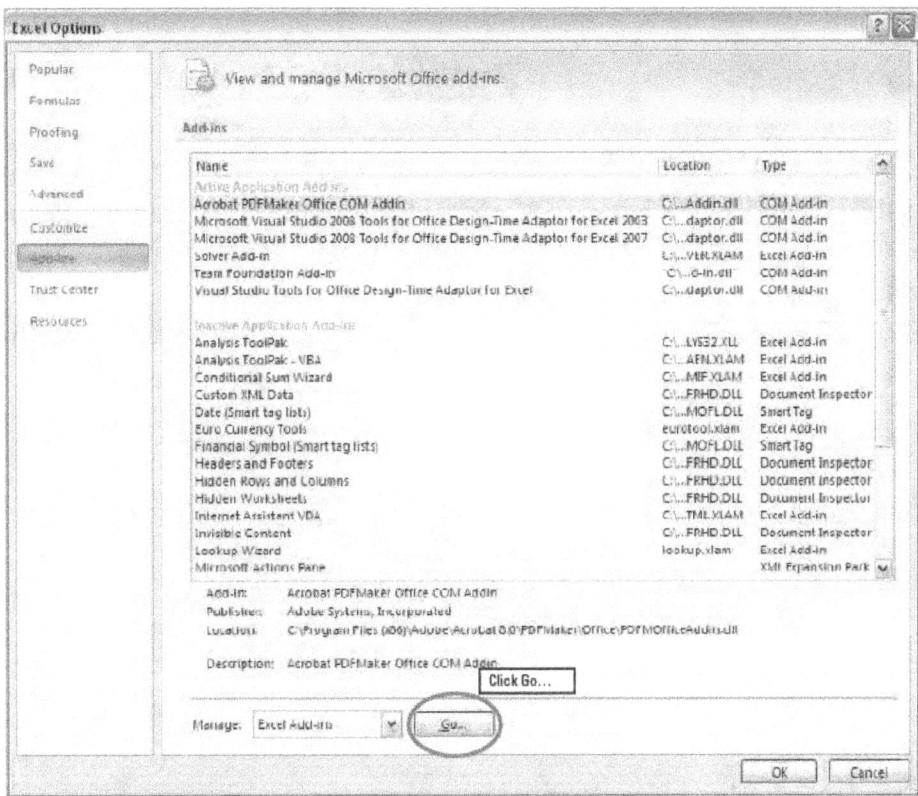

Figure B4.   Screenshot of the Excel Options dialog menu.

Figure B5.   Screenshot of the Add-Ins dialog box.

www.ingramcontent.com/pod-product-compliance
Lightning Source LLC
Chambersburg PA
CBHW081608170526
45166CB00009B/2876